DISCARDED

Zoos in Postmodernism

Zoos in Postmodernism

Signs and Simulation

Stephen Spotte

Madison • Teaneck
Fairleigh Dickinson University Press

© 2006 by Stephen Spotte

All rights reserved. Authorization to photocopy items for internal or personal use, or the internal or personal use of specific clients, is granted by the copyright owner, provided that a base fee of $10.00, plus eight cents per page, per copy is paid directly to the Copyright Clearance Center, 222 Rosewood Drive, Danvers, Massachusetts 01923. [0-8386-4094-X/06 $10.00 + 8¢ pp, pc.]

Associated University Presses
2010 Eastpark Boulevard
Cranbury, NJ 08512

The paper used in this publication meets the requirements of the American National Standard for Permanence of Paper for Printed Library Materials Z39.48-1984.

Library of Congress Cataloging-in-Publication Data

Spotte, Stephen H.
   Zoos in postmodernism : signs and simulation / Stephen Spotte.
     p. cm.
   Includes bibliographical references.
   ISBN 0-8386-4094-X (alk. paper)
   1. Zoos—Philosophy. 2. Postmodernism. I. Title.
   QL76.S66 2006
   590.73—dc22

2005019098

PRINTED IN THE UNITED STATES OF AMERICA

How did he elucidate the mystery of an invisible person, his wife Marion (Molly) Bloom, denoted by a visible splendid sign, a lamp?

—James Joyce, *Ulysses*

Contents

Acknowledgments	9
Introduction	13
Chapter 1	22
Chapter 2	37
Chapter 3	49
Chapter 4	66
Chapter 5	75
Chapter 6	91
Chapter 7	102
Chapter 8	124
Chapter 9	133
Chapter 10	142
Chapter 11	153
Notes	164
Literature Cited	179
Author Index	187
Subject Index	191

Acknowledgments

DR. R. BRUCE GILLIE REVIEWED THE MANUSCRIPT. STEVEN SHEPARD assisted with the graphics. Hana Amichai graciously allowed me to reproduce three lines of poetry written by her late husband, Yehuda Amichai, not yet included in any anthology of his work.

Zoos in Postmodernism

Introduction

ZOOS ARE FOR PEOPLE. WE GO THERE TO GAWK AT THE ANIMALS AND ponder, if briefly, what their world might be like. Some believe the reasons are deeper and more visceral, a palingenetic predisposition to associate with other living creatures no matter what the circumstances, recalling notions of a primitive yearning stamped onto the human psyche early in our evolution. This, we are told, explains both our attraction to animals (especially those with human characteristics) and the capacity to appreciate beauty in Nature. Such ideas are intriguing, but because they have yet to be molded into scientific hypotheses and tested empirically we must confine them to the realm of beliefs and look elsewhere for explanations.

In this essay I offer a strictly cultural interpretation of zoos as they exist today in postmodern society, particularly in the United States, and of our responses to captive animals. I argue that postmodern zoos—zoological parks, wild animal parks, menageries, and public aquariums—do not exist because such things are presently impossible. To create one would involve forcing it into a configuration similar to film, narrative fiction, or art, and were that to happen captive animals might then become expendable, replaced by images or simulacrums. The genre being what it is, there can only be modernist zoos in postmodern times, making cultural anachronisms of animal collections as we now know them.

Among postmodernism's distinguishing features is a ceaseless flood of texts and images devoid of reference or meaning, evidence to some of Marshall McLuhan's dictum that medium and message are indivisible, the message component requiring only "circular response, verification of the code."[1] Their conflation announced the end of both, including any link between different perceived realities.[2] Walter Benjamin anticipated the closing disparity when he wrote, "The replacement of the older narration by information, of information by sensation, reflects the increasing atrophy of experience."[3] Also in the guise of prophet, Charles Baudelaire warned how history and memory were in danger of being transformed by the photographic image.⁴ More recently, Fredric Jameson describes postmodernism as "the disappearance of history, the way in which our entire contemporary social system has little by little begun to lose its capacity to retain its own past, has begun to live in a perpetual present and in a perpetual change that obliterates traditions."[5]

Benjamin believed the press to be the event, not the messenger, creating "the illusion that deeds are reported before they are carried out."[6] Benjamin, a German Jew, killed himself in 1940, or twenty years before the beginning of postmodernism, but his prescience survives in many places, including Orhan Pamuk's novel *Snow*. A poet named Ka visits the office of a provincial Turkish newspaper. As the day's edition comes off the press he sees notice of a recital that evening at which he will read his new poem called "Snow." Ka turns to the editor and tells him that he has written no such poem. The editor replies, "Don't be so sure. There are those who despise us for writing the news before it happens. . . . And quite a few things do happen only because we've written them up first. This is what modern journalism is all about."[7] Prolepses of this sort could explain why the evening news always begins at six o'clock and lasts exactly thirty minutes.

Film, and more directly television, has altered our relationship to history and by extension to the natural world. As Anne Friedberg says, "The past is, now, inexorably bound with images of a constructed past: a confusing blur of 'simulated' and 'real.'"[8] Moreover, "The cinema developed as an apparatus that combined the 'mobile' with the 'virtual.' Hence, cinematic spectatorship changed, in unprecedented ways, concepts of the *present* and the *real*."[9]

Of course, present and real have been skewed since Plato's time. Virtual reality, postmodernism's supreme achievement, merely highlights the recondite nature of what philosophers and scientists who study perception call *naïve realism*—our commonsense view of the world. By their interpretation the "realism" we see is probably illusory, a system of representations invented by our sense organs and central nervous system.[10] Whether "real" or virtual we adjust easily to streaming images, evidence of a refined perceptual entelechy that might even confer a hesitant evolutionary fitness. Just as the tussling of young lions can be seen as practice for the adult hunt, video games are a child's preparation for future cybernetic conferences and missile launches. Michael L. Benedikt reminds us that in the virtual world postmodernism is never turned off whether or not we as individuals participate, although the warning is clear: "Games you cannot afford to leave are not games."[11]

Today's zoo can't escape certain connotations, notably restricted space, unpleasant odors, and the uneasy truth of humankind's dominance over wild creatures. But in a culture where reality and image have become synonyms, captivity merges easily with consumerism and in the ensuing metamorphosis loses many of its harsh trappings. The frailties and stress associated with zoo life then disappear along with our sympathy, and the animals are transformed into their own images. We see it in theme parks and in theme park/shopping mall hybrids: gambling casinos with dolphin shows and white tigers, shopping malls with petting zoos and rainforest exhibits, nightclubs decorated with aquariums.

Like movies and television, postmodern shopping malls hold time in abeyance, expand space, and ablate the weather. Objects are made over as images and propelled into virtuality, including any animals kept for the purpose of indirect commodification. The merging of theme park and mall is the closest any derivative of the conventional zoo comes to immediately escaping its modernist heritage. Future zoos might exhibit only animals bred in captivity for many generations, perhaps remnants of species that were once endangered and finally extinct in the wild. These remaining specimens, the artifactual progeny of selective breeding and genetic tinkering, will be postmodern in every way. The results should please those who show the proper historicist reverence for fragments and a romanticized belief that the past can be reconstituted.

Visual repetition is a means of sustaining the present, the equivalent of chanting words or phrases, as in incantations (fig. 1). To Jameson the fracturing of time into a series of repetitive presents is one condition of postmodernism, the other being the transformation of reality into images.[12] Two additional identifying features are social pastiche and schizophrenia, and an overpowering element that contemporary critics have called "death of the subject," or, in Jameson's words, "the end of individualism as such."[13] Individual style, readily identifiable in modernism, has dissipated in a flood of sameness driven by societal dementia and the homogenization of cultures. Benjamin anticipated this when he wrote, "Novelty is a quality independent of the intrinsic value of the commodity."[14] As I hope to show, nowhere is conformity more apparent than in zoos and the methods by which they transmit information. And what about this

Fig. 1 *Twenty Marilyns.* (Andy Warhol, 1962) © 2005 Andy Warhol Foundation for the Visual Arts/Artists Rights Society (ARS), New York; TM 2005 Marilyn Monroe, LLC by CMG Worldwide, Inc./www.MarilynMonroe.com.

information? Can we say that captive animals themselves impart knowledge of wildness or is the association too chipped and faded for our postmodern perception? Theodore Link questioned this relationship in 1883, and his points seem valid today:

> I have simply found that an animal, as closely confined as most of them are in zoölogical gardens, retains *none* of its *natural habits;* it only exists—a mere automaton; and even this existence is seemingly under protest.

Therefore, this aforesaid "study and dissemination of knowledge, etc.," is "a delusion and a snare."[15]

Recognizing dimly the power of images to obliterate memory, zoos in postmodern times define themselves as leaders in conservation research—static Edens—and as "arks" in which Earth's endangered species can safely be sheltered. They count on the spectator's amnesic gaze and the knowledge that nobody cares that zoos were once—and will always be—collections of curiosities. One of my hypotheses is that animals held captive in these facilities have relinquished their ontological status as part of the natural world. Unable to reproduce as they might in Nature, to expand their ranges, and to evolve, they can only simulate their wild conspecifics. Now reduced to static objects, we then view them as spectacles.

The zoo today is a cultural fossil, a barren evolutionary branch in the leafing of postmodernism. But it was not always so. Early on, a weak case could be made that here indeed was actual Nature, if not in the raw at least very close. Earlier zoo-goers saw the lion's fearsome teeth, felt the water buffalo's power. There was awe when the caged vulture spread its wings. Such attributes seemed real, the animals their embodiment. All it took was naïveté and a misdirected imagination after the manner of Flaubert's fictional characters Bouvard and Pécuchet. Today's zoo occupies a netherworld between real Nature, which can never be known, and a leaky bucket of cinematic "realism" that retains no secrets at all. Zoos claim to represent the first while competing actively with the second. Grasping at both, they fail to be either.

Postmodernism is also defined by the *hyperreal*, models of reality without origin that by their appearance cause the objects they represent to disappear (fig. 2).[16] Its most important expression is the production of *simulacrums*, new realities invoked as believable copies but devoid of meaning or reference. In hyperreality, which is neither true nor false, models of the real become substituted for reality itself, which is then beyond reach. The entire process is driven by *signs*—mental constructions of reality and memory—that in postmodern times dissipate any distinction between immanence and image. Signs in the form of pictures blend time and culture, becoming more real than the objects they represent (fig. 3). Information increases relentlessly, but with a proportionate loss of meaning. Jean Baudrillard says, "It is all of metaphysics that is lost. No more mirror of being and appearances, of the real and its concept. No more imaginary co-

Fig. 2 Bare-breasted Indian maiden painted on a steer skull, icon of the American southwest. Santa Fe, New Mexico. (Stephen Spotte)

Fig. 3 Ancient Mayan civilization in the postmodern age. Yucatán, México. (Stephen Spotte)

extensivity: it is genetic miniaturization that is the dimension of simulation."[17]

We ingest the hyperreal as genetically modified plants and animals, incorporating their constituents into our postmodern selves made beautiful and timeless by cosmetic surgery and prostheses. Cryogenic sleep awaits while we dream restless dreams of being cloned and touching immortality. We recognize vaguely that Dolly the sheep ate hay and defecated on the ground but existed in another realm. Information trapped in silicon now resides in cyberspace where humanity's collective history is sorted, organized, and distributed. Imagination has become stereotyped, computer games fog the mirror of existence, and everything can be reproduced infinitely.

Worst of all, references to reality have been replaced by a system of signs more insidious and adaptable than meaning. This secondary incarnation, although entirely artificial, has replaced the real. Life, it seems, is just like a movie. Even *Nature,* or its synonym *the Other* (so-called because it represents the antithesis of human consciousness), has shrunk to a system of signs disguised in images. Indeed, signs *have become* reality—hyperreality.

If, as Baudrillard predicts, we are now immersed in the hyperreal where spatial perspective ceases to exist, then most of the *spectacular*—that is, the witnessing of objects or events by spectators—will not survive, including modernist zoos.[18] As postmodernists we seek to erase dimensional space by displacing it with ourselves until we blend fitfully into the dyslexia and asperity of the times. Television dooms us "not to invasion, to pressure, to violence and blackmail by the media and the models, but to their induction, to their infiltration, to their illegible violence."[19] At football games, ostensibly spectacles, fans wear odd costumes and paint their faces in an effort to *become* the event. This doesn't happen at the zoo. As I shall argue, going to the zoo is strictly spectacular: zoo-goers are passive watchers. This places the zoo squarely within the modernist era, generally agreed to have ended with the 1950s.[20]

So what? you ask. Is this so bad? After all, if postmodernism represents cultural degradation, a time when society loses touch with itself and its heritage, then maybe zoos had best remain entrenched in the previous era. I might agree if modernist zoos stood for something worth keeping. As I shall demonstrate, even in early modernism zoos lagged behind other protocinematic activities. Well before the modern era both the panorama and diorama were technologically and esthetically more advanced than any animal display. Despite self-serving arguments to the contrary, zoos have done little to distinguish themselves ever since. Finally, postmodernism is here, like it or not, and zoos must adapt or become as endangered as the species they purport to save.

Before offering an unorthodox view the critic customarily prefaces his argument by assessing the status quo as a prelude to refuting it. But in this case there is no orderly thesis to refute.[21] It would seem that zoos have always existed beneath the radar of theory and philosophy, held captive themselves by an unctuous form of consumerism. Caged birds can't soar.

Zoos point to their numerous programs of conservation, science, and education, but noise alone is insufficient to muffle doubt. The fact is, those who administer them have been disingenuous. Zoos breed endangered species selectively, choosing mainly high-profile vertebrates that visitors are likely to recognize or might pay to see. This offers an image of Nature no less skewed than cartoonist Saul Steinberg's 1976 *New Yorker* cover, a map titled "View of the World from 9th Avenue," in which countries and continents—everything west of the Hudson River—appear as mere smudges on the horizon. The most touted advances to which zoos can point (e.g., artificial insemination, the maintenance of frozen tissue banks) are technological, not scientific, and zoos are not the only facilities equipped to use them. Zoos are woefully thin on original science, and their societal value in education, which has yet to be measured properly, remains undemonstrated.[22]

Zoos are easy to criticize but difficult to critique, partly because they seem driven by aleatory swings in public perception. This has led to a bunker mentality and failure to assume any honest responsibility for their own philosophical deconstruction. In the absence of a formal theory I could find no means by which to investigate the zoo's true nature, the depth of its metaphysics. Instead, I looked to theories of literature, film, art, photography, and science. Zoos appear to me like unwieldy appendages on these other disciplines, parasitizing them as convenience and opportunity allow but unable to bring themselves into focus. They seem to comprise neither art nor artistic expression, science nor the scientific institution, conservation nor "centers of conservation."

Efforts have been expended intermittently over the last 150 years to make American zoo exhibits more "naturalistic." What this entails is adding simulated rocks, trees, tree stumps, plantings (alive and simulated), streams, grassy patches, and so forth. We may *think* this has some positive effect on the inhabitants, but seldom is there confirming evidence. In truth, beyond a few generalities (i.e., otters and ducks are swimmers, giraffes are browsers) we know little about which elements in the natural habitat are most important and which only seem important because they appear so to us. Nor do we know how these elements interact even in Nature. When it comes to identifying Eden's components, postmodern humans are scarcely more knowledgeable than Adam and Eve.

Zoos are neither natural nor naturalistic. In the very act of confining wildlife they establish unnatural settings and are themselves unnatural by definition. As with illustrations of unicorns, degrees of "naturalness" in zoo exhibits have no basis in reality. A sterile cage with bars is just as "natural" or "unnatural" as a barless exhibit decorated with rocks and trees, either real or simulated. Then what are zoos? Simple spectacle, I shall argue, destinations of amusement. And although zoos sometimes employ art, film, photography, technology, and other well defined disciplines, they nonetheless remain apart. Their aura, like a pile of raked leaves, seems at once familiar and insubstantial.

My place in this essay is not to criticize how zoos are managed or whether keeping animals in captivity is ethical. These subjects require a different treatment and have been addressed by others. I intend to argue that zoos are not what they claim to be—loci of conservation, science, and education—but rather isolated islands of simulacrums and confusing semiotic signs where visitors are spectators and the animals passive curiosities. I shall argue, in other words, that zoos have advanced little heuristically since the beginning of the modern age. Whatever originality my thoughts contain is modest. We cite others or quote them in the interest of scholarship, but ideas are timeless. As Karl R. Popper reminds us, there is no ultimate source of knowledge, and the search for it leads only to infinite regress. Moreover, Benjamin writes, "To convince is to conquer without conception."[23] Is the argument that follows appealing conceptually, merely convincing, or neither? You decide.

1

> Among the many certainties whose lack he complained of, one alone is present, and it is that all things appear to us as they appear to us, and it is impossible for them to appear otherwise.
> —Umberto Eco, *The Island of the Day Before*

WHAT'S REAL? DOES THE ANSWER MATTER? IGNORE THE NOTION OF Plato's *Forms*, his positing of eternal and changeless abstract objects of the *eidos*. The reality I discuss here is not classical philosophy but something less ethereal, the bent light of our worldly conceptions forced through a postmodern prism. Where zoos are the subject the task is best accomplished by dissecting a hypothetical exhibit using semiotics and then examining the various limbs and organs as a test of articulation. I shall get to it after describing the tools used in the demonstration.

It was 1928 or 1929, the year is uncertain, that Belgian Surrealist René Magritte painted a picture of a pipe (fig. 4), a common male accessory of the times. The object is ordinary—a simple bowl with a curved stem—and so is its representation. In fact, there would be nothing to distinguish Magritte's painting at all were it not for some curious text. Underneath the image Magritte wrote, "This is not a pipe." By this innocuous statement he highlighted the apposite truth that representation and reality are not the same. Here I extend the notion of Magritte's pipe to zoo animals in an effort to cast them metaphysically in the light of postmodernism. In the epigraph above, semiotician and novelist Umberto Eco provides a teasing glimpse of reality. *All things appear to us as they appear to us,* he says, which is quite different than if he had said, *All things appear to us as they are.* Like Magritte's pipe, all things indeed appear to us, but never as they are.

A zoo exhibit with its attendant species labels and graphics presents two problems of interpretation, one connotative, the other ontological. Both impinge directly on meaning and the representation of reality. In particular, the intentional images offered zoo visitors—those containing information considered desirable—must overcome extraneous visual, acoustic, and tactile noise in the form of

Fig. 4 *This is Not a Pipe.* (René Magritte, 1928–1929) © 2005 C. Herscovici, Brussels/Artists Rights Society (ARS), New York.

advertising, logos, lettering on the tee-shirts of passers-by, screaming children, chewing gum congealing on sidewalks, and even a zoo's own information, some of which competes directly with the intended message.

The hypothetical effectiveness of a zoo exhibit can be modeled using *semiotics*, the study of signs—not just zoo graphics and road signs, although these too can be included—but everything we take to represent something else and imbue it with meaning.[1] Language is a semiotic system in which words give rise to mental images representing objects or concepts with meaning. All human knowledge takes the form of signs; to others, we ourselves are signs. Even sensory information is received in signs, none of which is directly real. C. K. Ogden and I. A. Richards write: "The sensations which lie at the basis of all perceptions are subjective signs of external objects. The qualities of sensations are not the qualities of objects. Signs are not pictures of reality."[2]

In the following discussion I use semiotics to identify, organize, and test abstract relationships. As a scientist I take comfort in this approach. The humanities seem less frightening when pieces can be extracted here and there and relegated to testable hypotheses. Like water, ideas flow strongest and fastest in a restricted channel. The

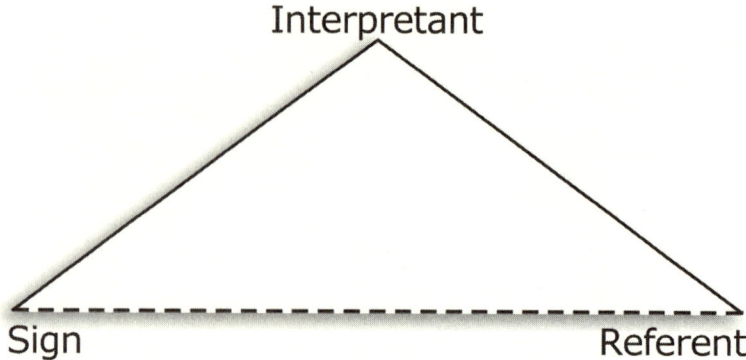

Fig. 5 The semiotic triangle. Redrawn from various sources.

semiotic process can be shown diagrammatically as a triangle (fig. 5). At one corner is the *sign* (also called *signifier*, or *representamen*), usually conceived as a material object or sensory stimulus.[3] Typical signs are the odor of coffee brewing, the sight of a flashing traffic light, a road sign, the feel of sun on your back, and the sound of raindrops hitting a window.

The sign is linked directly to the *interpretant*, or *signified*, at an adjacent corner of the triangle. Consider the interpretant as a thought representing the sense made of the sign (i.e., how the receiver interprets it). Think of it also as the *meaning* obtained from the sign. Someone smelling coffee perhaps unconsciously interprets the odor as a sign of coffee brewing nearby; someone hearing raindrops hitting the window does not have to glance up to know it's raining.

At the remaining corner of the triangle is the *referent*, to which the interpretant both refers and serves as mediating agent with the sign. The referent can be indirect (a memory of drinking coffee) or direct (sight of a coffee bar on the corner). Other examples are suddenly being cold upon hearing rain, or glancing at an umbrella in the corner. In other words, there arises a connection at least vaguely familiar from having been experienced in some fashion. A person who has never been sunburned could, on feeling the sun on her back, recall knowledge of its effect gained from reading, previous training in dermatology, or hearing about a friend's experience. Note by the dotted line in figure 5 the absence of a direct connection between sign and referent. This means that words (if words are the pertinent signs) are neither parts of things nor do they consistently imply things with which we make them correspond. The most we can do is *impute* a relationship between sign and referent.[4] Also keep in mind that a sign must be present before what it represents can be ascertained, but this is impossible without the interpretant's mediating action.

Suppose that while walking past an outdoor fruit stand you glance at a stack of round objects. You note their size and orange color. Instantly interpreting these images (in sum, the *sign*) as characteristic of an edible fruit unlike lemons, limes, grapefruits, and apples, you recognize them as oranges (*interpretant*). The sign until interpreted is incomplete, an unconnected fragment (e.g., size and orange color). We interpret a sign as *signifying something other than itself*. Interpretation of the sign in this example brings to mind the taste of an orange, the feel of its skin, how it comes apart in sections and makes your fingers sticky (*referent*). The *interpretant* has mediated between the *sign* (a glimpse of orange objects of a certain size) and the *referent* (knowledge of oranges based on previous experiences of having eaten them). To someone who had never seen or tasted an orange the initial sign of a stack of cylindrical orange objects would have no meaning, and the *signification* (i.e., the connection between the sign and its interpretant) could never be completed.

Some signs are generic enough to be classified. With *symbols* the association between sign and interpretant is arbitrary and must be learned. As Thomas A. Sebeok tells us, "A *symbol* is a sign that stands for its referent in an arbitrary, conventional way."[5] Languages (including grammar and punctuation), mathematics, navigational aids, and national flags are symbolic systems. According to Sebeok, "An *icon* is a sign that is made to resemble, simulate, or reproduce its referent in some way."[6] All pictures are iconic, including cartoons. So are metaphors and the depiction of iconic figures of men and women on the doors of public restrooms. Onomatopoeic words are iconic signs because their referents are mimicked acoustically. Scented floor cleaners that smell like pine or lemon oil are iconic of the odors of pines and lemons. With an *index*, sign and interpretant are directly associated: smoke/fire, thermometer/temperature, rash and itching/poison ivy, lightning/thunder, scream/fear. Sebeok notes: "The most typical manifestation of indexicality is the pointing *index* finger, which humans the world over use instinctively to point and locate things, people, and events in the world."[7] Signs without any basis in reality can also be iconic.[8] Eco remarks, "The image of a unicorn is not similar to a 'real' unicorn; neither is recognized because of our experience of 'real' unicorns, but has the same features displayed by the definition elaborated by a given culture within a specific content system."[9]

A sign can sometimes be both symbol and icon, and even an index. The stylized ungulate skull used in American zoo graphics to represent endangered species is usually considered a symbol, although its iconic function is stronger because, like the Christian cross, it "can be subjected to manipulations of the expressions which affect the content."[10] The skull is also indexical: skull/death. To summarize, Mickey Mouse is a symbol of Disney, but an icon for mice; his droppings and urine trails (if any) are indexical.

Picture a fenced enclosure containing a single animal, otherwise empty except for several small trees with their trunks wrapped in wire and a pile of large rocks stacked haphazardly. A species label identifies the animal on view as a specimen of Alpine ibex, *Capra ibex,* and a shadow illustration highlights the species' phenotypic characters: flat horns bearing flat ridges, and a small beard. All species of ibexes, we are told, are goats. A nearby graphic illustrates, by means of a shaded map, that the Alpine ibex was once distributed across the European Alps, but its range today is much reduced. A nearby photograph, backlit to emphasize its colors and high resolution, depicts a small group of wild Alpine ibexes standing on a high snow-covered crag. Additional information tells us that the animal in this exhibit is a male named Hans, that he was born in captivity on May 23, 1998. Hans' principal foods in nature would be grasses and forbs, but here at the zoo he is fed vegetables, grains, and hay. The gestation period of *C. ibex* averages twenty-three to twenty-five weeks. Young Alpine ibexes travel with their mothers in maternal herds of ten to twenty.

Hans is obviously alive, a condition that by no means affects his actual status. He moves around his enclosure, eats, drinks, defecates, maintains a stable body temperature. His vision, hearing, and other senses are sharp, and he responds promptly to external stimuli. At first glance there seems no doubt that Hans is a *real* ibex. But he isn't. We could say, in the language of Peircean philosophy, that the ibex in this exhibit represents a *semiotically real object* incapable of linking sign and referent (fig. 6).

The species label is incidental information mostly devoid of insight, but the material in the graphics (in particular, the diagnostic species characters) comprises another kind of image, a sign. Not surprisingly, the semiotically real object determines its sign, or the reverse of what the graphics imply. To state this differently, the shape

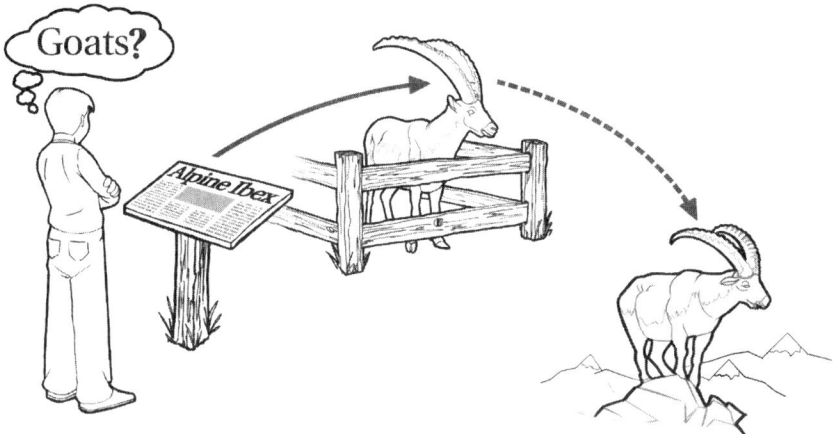

Fig. 6 Semiotic breakdown of an Alpine ibex exhibit. Spectator reads a graphic (*sign*). He thinks about goats (*interpretant*) and sees the ibex on display (*semiotically real object*). The actual *referent* exists only in Nature. (Stephen Spotte)

of the horns don't determine this species of ibex; rather, the reverse is true.

The interpretant in a semiotic process is context-dependent, which naturally affects how a sign is linked to its referent. Charles Sanders Peirce wrote that the interpretant comprises "all the facts known about its object."[11] To someone having extensive knowledge of goats, the graphic materials release an unconscious compendium of facts and mental images acquired through reading, experience, or both. This corpus of knowledge is the interpretant, or the "conceptual ibex," more accurately, members of the family Bovidae.

However, signification ends abruptly for zoo spectators completely ignorant of goats, not an unusual situation in urban America where the only goats likely to be encountered are in petting zoos or on TV. Under such circumstances the formation of interpretants, which depends on context-dependent relationships, is impossible. Peirce noted the underpinnings of this dilemma, discussing how it would still be possible to link a sign to its referent when previous experience of it was required to trigger any recognition. Being acquainted with the natural sciences he used lithium as an example, stating that lithium "dissolves when triturated." Thus the word "lithium" *denotes* its object by describing a method that allows it to be identified.[12] Although well-grounded epistemologically, such a cumbersome process does little to solve the immediate problem of a casual zoo spectator, which is insufficient experience to form appropriate interpretants.

Eco treated the subject of unfamiliar signs extensively, tracing back to Kant's *schemata*. Judgment, Kant had reasoned, is the facility of the particular as a part of the general. If the general is known (in the form of a law or rule), judgment is *determinant*. However, if the particular has been presented and the general must be sought, judgment is *reflective*. If we recognize an ibex to be part of the larger class of goats (particular → general), our judgment is determinant (i.e., knowledge of the particular fixes the outcome of the general by restricting choices). If the ibex is *not* recognized to be a goat (that is, its inclusive class, Bovidae), judgment must be sought and is reflective, or indeterminate, until the answer has been found. What Peirce called *abduction* (a sort of feel, or innate knowledge, for what is correct) is synonymous with reflective judgment.[13]

Following the Kant/Eco model, we would expect the zoo spectator to intuitively "know" that ibexes are goats. In my opinion, this is far-fetched. Stated differently, when seeing an ibex for the first time a knowledgeable spectator would recognize it as a *token* (ibex) of a *type* (goat). A token stands for, or is representative of, its type, but if the type is unknown, how can the spectator using abduction (or reflective judgment) arrive at an answer when the very rules (taxonomy in general, taxonomy of the ungulates in particular) are a mystery? Clearly, an impossible situation. Why? Because the route to solving the problem is circular. Eco writes: "We recognize individuals because we relate them to a type, but we are able to formulate types because we have experience of individuals."[14]

I see still another problem. In the study and subsequent understanding of Nature, verisimilitude is in the details. Deprived of subtle images and objects that place our Alpine ibex in context—and unable to recognize them even were they present—a casual spectator's only hope is to form a representational mental construct of his own. The rock pile, species label, and color transparency have been placed there for the purpose of stimulating such a representation, but this is a false hope in the absence of experience.

Suppose the graphics include visual aids, perhaps a phylogenetic tree or cladistic dendrogram illustrating the kinship of ibexes to the other goats and of the goats to all extant ungulates. How is this information useful? We could, I suppose, model perceptual characteristics of a "typical" zoo spectator's brain, what Eco calls the "black box," but such an approach is unlikely to lead anywhere. "Black boxes," by definition, are completely sealed, and their contents can be tested only by inference. I shall use instead Eco's cognitive models, starting with his *cognitive type* (CT).

Someone who has never seen an ibex can nonetheless assign to it some general characteristics: hairy, eats hay and vegetables, four legs and hooves, curved horns, bearded. Certainly some sort of goat/sheep/deer/antelope. Obviously not a cow or horse, probably not a moose, and certainly not a tree. To a reader of children's literature the beard might be a clue (don't billy goats have beards?). This mental process is similar to a Kantian *schema* except for lack of a bridge linking concept with intuition. The spectator, having never seen an ibex and ignorant of taxonomy, has neither conceptual nor intuitive knowledge at hand. She does not, in other words, have a CT for ibexes, recognizing them with certainty only as animals. There are the horns, of course, but these lack the indelible quality of an elephant's trunk, a lion's mane, or a zebra's stripes. For James Joyce the matter of animal indelibility was just such distilled essence, although the factors triggering recognition could be far more subtle and inclusive. In *Ulysses*, Joyce tells us that "Horseness is the whatness of allhorse."[15] A horse, in other words, is a composite of traits encompassing nothing more nor less than everything horsy.[16]

The graphics explain how to distinguish goats from their closest relatives the sheep, and visual taxonomic aids illustrate this and other aspects diagrammatically. The information is saying, in effect, you have just seen the particular, or token (ibex), and what you are now reading stands in for the general, or type (goat). But no graphic can form an interpretant of that hazy quality called "goathood"—a CT based on experience with individual goats of one species or another. Unfortunately, "goat" does not connote a semiotic "primitive" triggering an instant CT as, for example, flight/bird or fish/swim. Unless our spectator is a trained zoologist or has extensive knowledge of goats, type will remain elusive, and comparing tokens would be an empty exercise.

As Eco tells us, naming is a social act enabling members of a society to recognize various individuals (species in this case) as tokens of the same type.[17] Token and type in a zoo exhibit have no relevance beyond casual interest. Common pets are different. When referring to a pet we might tell a new neighbor, *This is Fido.* The neighbor might reply, *What a handsome dog!* or *What a handsome specimen of Canis familiaris!* A Korean neighbor could exclaim, *What potentially delicious suyuk!* (boiled dog meat with spices). Regardless of context, Fido has been correctly identified as a token, or individual (Fido, *Canis familiaris, suyuk*), of a type (dog, the genus *Canis*, meat dish). Although Fido's several denoted names identify him to different sections of society, in all cases the CT—the mental registering of a sign somehow representing *dog*—is achieved effortlessly because of previous experience.

The graphics near our ibex exhibit are similar to Eco's *nuclear content* (NC), which he differentiates from *meaning*. Meaning is ordinarily associated with a private mental experience. Nuclear content, in contrast, is a public expression representing a communicative consensus about which guidelines can be used to identify a token of the type.[18] In doing so, NC acts as a cognitive compass, steering toward the formation of a hesitant CT. By being presented publicly zoo graphics qualify, and also because they represent a communicative consensus of two types. First, they are written in English, our dominant language in the United States. Second, the information conforms to general guidelines and traditions established by the closed community of zoo administrators.

Eco emphasizes that unlike the CT, the NC can be seen and touched. Ibexes presented as symbols, icons, or indices (that is, as signs) fall under the rubric of NC if they provide a way of identifying a token of a type (i.e., help identify the referent). Importantly, the CT can be transmitted within a society by means of the NC. In the case of zoo graphics someone approaching with no cognitive type in mind (i.e., no mental notion of goats) can presumably develop one from the NC, or after studying the graphic materials. The association, according to Eco: "We postulate a CT as a disposition to produce an NC, and we treat an NC as proof that there is a CT around somewhere."[19]

Complex knowledge is possible in the presence of adequate information. Eco calls this knowledge *molar content* (MC). Complex knowledge encompasses perceptual recognition of a token based on experience and might include the vocalization, gait, silhouette, or resting posture of Alpine ibexes, but also knowledge of their daily food intake, types of foods eaten in the wild, gestation period, or taxonomic position; in other words, at least some of the encyclopedic information in the graphics. The presence of MC in a zoo graphic is proof of its intended function to impart complex knowledge. Whether this actually happens is unknown.[20] Still, the orderly acquisition of complex knowledge in the absence of experience seems improbable.

Why must the interpretant in this example include *all* goats and not just the species *C. ibex*? Because by providing distinguishing characters of the token the graphic necessarily admits the type. Taxonomic characters are useful only by subtraction, pertaining to a given species after those defining its closest relatives have been eliminated. Was opening the door for every other goat, even tacitly, a good idea? That depends on the motives of the zoo's administrators. They had no alternative if their objective was to show *C. ibex* as a member of the ontological class of goats. However, if the purpose was simply to explain something about the Alpine ibex then including information about its relatives seems irrelevant and potentially confusing. To properly place *C. ibex* in context (i.e., within the family Bovidae) the spectator must be capable of envisioning, by means of the interpretant, one goat after another until *C. ibex* comes into view distinguished from the others by its unique phenotypic characters.

A zoo would not be shirking its duties by omitting taxonomic information. After all, zoological taxonomies are not necessary for recognizing the referent. In general terms, the ability to identify and then classify a representation is independent of any prior knowledge needed to classify its object.[21] As Eco reminds us, "no one has ever denied that someone is capable of perceiving and recognizing a platypus without necessarily knowing whether it is a Mammal, Bird, or Amphibian."[22] Such categorization is arbitrary, as Kant recognized when dropping taxonomy into the pool of reflective judgment. Knowing that your wife is a specimen of *Homo sapiens* is of little use when searching for her in a crowd. Similarly, complex knowledge is of limited help except to the expert. In fact, encyclopedic information such as zoos offer is superfluous where it impinges on recognition. No matter how extensive your conceptual and intuitive knowledge of human anatomy, evolution, and behavior, finding your wife among all those people requires quite a different *gestalt*. Critics might complain that a man's recognition of his wife is somehow irrelevant to my argument because of her individuality. However, as I discuss in chapter 2, Hans the Alpine ibex is an individual too, and so is *C. ibex*, his species.[23] Call him Hans or a specimen of *C. ibex*, it makes no difference. Standing among sheep or ponies any school child would pick him out easily, just as Hans' keepers could distinguish him in a group of his conspecifics.

Note from figure 6 that the *real* ibex—that is, an Alpine ibex living in the European Alps—is remote, connected tenuously by dashed lines to the rest of the scheme, and connected to the spectator not at all. Looking at the figure we might take the graphically short distance between semiotic and real at face value, but in fact the dashed line stretches well past any relational horizon. The real ibex remains always outside our perception and understanding, tethered precariously to the object before us. For the sign-referent link to be tightened, Alpine ibexes would have to become extinct in the wild, at which point the captive ibex becomes the referent.

The *real* ibex exists in the wild and nowhere else. From our location at the ibex exhibit its sign is dormant and thus unexpressed. To choose an analogy, I imagine myself walking through a forest and encountering a fallen tree blocking the path. The tree is large, and its demise was no doubt a noisy event, but I can only surmise this having not been present. Suppose no one was there. Was the event still real? Of course, although not in a semiotically active sense; that is, no observer heard the sound (sign), conceived of a tree falling (interpretant), and connected the sign with the event somehow experienced, if only vicariously (referent). According to Floyd Merrell,

> I have no doubt that when a tree falls, it "really" falls, and with a crash, independently of any observer: the falling tree *is,* of course, what it *is.* If no observer is present, then the tree as a Peircean sign, as something representing [repeating itself] something to someone in some respect or capacity, has not been properly actualized (put to use). The physical event was "real" nonetheless, and the potential of a semiotic event for an agent, human or otherwise, was there; the potential for the sign's becoming authentic simply remained dormant.[24]

Similarly, Alpine ibexes in the wild are real, but their existence as signs has not been—and logically can't be—actualized by a zoo spectator observing a captive ibex, just as observing a tree standing in your yard fails to actualize the semiotic process of another tree of the same species falling over at that moment somewhere in a different country.

The captive animal before us can only be semiotically real.[25] Besides giving us a species name, which itself is arbitrary and thus problematical, a species label showing how to identify an Alpine ibex is redundant, merely overlapping the object itself while adding little else. The spectator's understanding of this same animal—that is, its signification—consists of mediating between label and semiotically real object, effectively conjuring a mental image of this, and only this, specimen. Meanwhile, the real ibex remains distant, invisible to zoo spectators and vested with no semiotic reality.[26] Whatever it is that a zoo exhibit reveals, it can tell us nothing about Nature.

The color photograph meant to juxtapose the captive ibex with a wild conspecific in its natural habitat serves as a duplicate object no different semiotically than the captive ibex itself. This is because the real ibex can never be known; simultaneously, its photographic image lies trapped in iconism. Although some of the graphic materials are undoubtedly attractive and meant to be didactic, they appear to serve no useful function. Tautological or irrelevant information also includes the rock pile, phenotypic characters on the species label, and any other material not directly referring to the semiotically real object and therefore outside this specific system of signs.

Keep in mind that the separate components of a zoo exhibit make up *strings* of semiotic signs. Because the management staffs of most zoos are part of a larger community of zoological institutions, the information presented must be acceptable to the body as a whole. Furthermore, this information, once approved either explicitly or implicitly (the latter by not being criticized), becomes societal law constituting the opinions and beliefs (*doxa*) of the closed zoo management community. What the spectator sees is the presentation of a semiotic system—live specimen, species labels, accompanying graphics and text—all intended to serve a heuristic purpose.

Undoubtedly, the most important object in an ibex exhibit is the ibex, and any distraction lessens its impact. But ibexes require explaining. Simply putting the spectator within view of one is insufficient if education is the motive. The typical zoo exhibit presents a semiotically real object reinforced with graphics purporting to tell about the real. If we assume that people visit zoos to view animals we can further assume any interest Hans generates is directed at him, not at the ontological class of Alpine ibexes. In other words, the spectators will be interested in Hans, a semiotically real object, rather than real ibexes (paradoxically, those existing only in the abstract). If so, consistency might be served by telling Hans' story, which is semiotically real too, rather than the hypothetical story of real ibexes. Alternatively, Hans' ontological status affords the perfect chance to tell a story about real ibexes in myth or legend. This is perfectly acceptable because Nature "out there"—the Other—can never be known to us anyway.[27] Stories, as I hope to show later, strongly influence our interpretation of the world.

Goats eat trees, and the trees in this exhibit have been wrapped with wire mesh to protect them. The wire wrapping serves as a semiotic sign, but signs are important only if they have meaning. Meaning presupposes understanding. Perhaps a label could be produced that reads *These Trees Are For Shade,* or *These Trees Serve An Esthetic Purpose,* or *These Trees Are Wrapped So Hans Won't Eat Them,* but doing so might detract from Hans. However, the wire wrapping itself is a distraction, and the exhibit might be better served if the trees were removed.

The rock pile neither symbolizes nor connotes anything obvious, nor does the ibex's presence necessarily provide context. If the rock pile is meant to be symbolic of a mountain, success is not assured:

sign (rock pile) → *interpretant* (presumed mental image of a mountain) → *referent* (presumed mental image of an ibex on a mountain)

Unless the animal is actually standing on the rock pile there is no context, and the probability of an associated referent is then much reduced. And unless the spectator can visualize the rock pile as a mountain and picture an ibex on it there can be no association at all. The referent lacks meaning by existing outside the spectator's experience. A high crag in the European Alps can *symbolize* wild nature, and an ibex standing at its edge might easily be seen as a *symbol* of freedom and daring. Stated differently, the crag *connotes* wild Nature, and the ibex standing there *connotes* freedom and daring. In *connotation* a sign stands for other signs associated with it.

Suppose the designer of this exhibit had more in mind than esthetics. Perhaps she placed the photograph nearby to catalyze the signification of sign → interpretant. Taking a cue from chemistry, catalysis seems an appropriate term, implying the initiation of a thought process (an association in this case) in which the photograph catalyzes the mental transformation of rock pile into mountain while keeping separate the original image of the mountain. Interpreted directly, the photograph is redundant if meant to reinforce any notion of rock pile as iconic mountain, and its presence and implied visual scale diminish the rock pile to what it actually is (a pile of rocks).

The distribution map, species label, and accompanying species characters are of doubtful value unless the spectator already has knowledge of the subject. Attributes of objects—including taxonomic characters—are beyond discovery except by interpretation. The inclusion of such information assumes that the semiotic process will be completed, culminating in understanding by the spectator. In this situation:

> *sign* (flattened horns bearing flat ridges + beard) → *interpretant* (memory of previous experience with goats) → *referent* (wild Alpine ibex)

Probably the most crucial corner of any semiotic triad is the interpretant. Suppose the spectator correctly equates the sign with the semiotically real object (captive Alpine ibex) but knows nothing at all about goats. Signs go unrecognized until interpreted, at which point they signify something other than themselves. In a large field containing goats of different kinds we might use the signs offered on the species label as field characters to distinguish an Alpine ibex from the rest. In this exhibit there are no others, and the task ought to be simple. To serve its function the interpretant must bring to mind some previous *experience* with goats, even memories of reading about them. To someone completely ignorant of goats the interpretant remains blank, the sign devoid of meaning. Without mediation through the interpretant the sign and referent stay unconnected, stalling the semiotic process until it finally fizzles out. Presenting species characters is obviously useless when only one species is present.

The name "Hans" is *denotative* by referring to, or denoting, a specific individual. In other words, as a sign the name *Hans* stands for something other than itself; it stands for a specific Alpine ibex. The birthdate is also denotative and in addition serves to reinforce Hans' status as a semiotically real object, identifying him as a narrative subject and principal protagonist in his own biography. By implication, Hans' story is representative of every Alpine ibex in captivity, just as any novel about an individual human being touches on the human condition. The stringing together of events forms the basis of narrative and also biography. Where wild ibexes are concerned the information is not partly irrelevant, but completely so.

Hans' foods are denotative in referring to him. Like his name and birthdate, what he eats reinforces his semiotic status and extends his biography: *This is Hans. He was born May 23, 1998. He eats carrots, cabbages, oats, and hay.*

The average gestation in the graphics speaks of the ontological *class* of Alpine ibexes and not *denoted individuals*. Stating how long Hans stayed in the womb would further reinforce his semiotic status and serve to continue the narrative. Instead of speaking of the generic ibex it might be more interesting to tell Hans' story—where he was born, his mother's name (further digressive denotation), and so forth, then state, *Hans was born after 167 days, a typical gestation time for Alpine ibexes.* Nothing is lost by doing this because the average, or mean, gestation is simply a number not pertaining to the semiotically real object at all.

Magritte's painting (fig. 4) requires revisiting. As Merrell states, "'This is a pipe' seems simple enough. It is a proposition (compound symbol) consisting of an index ('This') and an icon ('pipe') whose proper meaning entails at least tacit awareness of our 'dictionary' understanding of the word."[28]

If the image *is not* a pipe, several possibilities are apparent:[29]

1) This [pipe] (icon) *is not* a pipe.
2) This [image *of* a pipe] (icon) *is not* a pipe.
3) This [painting] (icon) *is not* a pipe.
4) This [sentence] (symbol) *is not* a pipe.
5) [This] this (index) *is not* a pipe.
6) [This] (index) *is not* a pipe.
7) [*Is not* a pipe] *is not* a pipe.

In terms of our exhibit we can safely state that the species label is not an Alpine ibex, nor is the photograph. Both contain images of ibexes and therefore are icons, equivalent to the second possibility above. What about Hans? Actually, a captive ibex as a semiotically real object is *represented* by the sign because, according to Peirce, a representamen, like any other sign, stands for something (or someone) in some respect or capacity. However, Hans is not the kind of sign we might call a symbol, nor is he an icon or index. If the Alpine ibex were endangered we could engage an artist to stylize Hans' outline into an image and use it as an icon of the species. Hans would then participate as a *dynamic object,* or the object that independent of the sign leads to the sign being produced.[30] The qualities of objects are different from the qualities of our sensations of them, and a sign is not an image of reality. As to the animal displayed in this hypothetical exhibit, perhaps only one completely accurate label is possible: *This Is Not An Ibex.*

2

> Is it possible to have a false perception of an illusion? Is there a true *déjà vu* and a false *déjà vu*?
> —Don DiLillo, *White Noise*

POSTMODERNISM IS A PANOPLY OF SIGNS SHAPED TO MIMIC THE REAL world and draped across it like Jorge Luis Borges' famous map.[1] This is the landscape of semiotic reality we occupy. Perhaps, in a reversal of Borges' story suggested by Baudrillard, the ground beneath us will continue to disintegrate until only the map remains.[2] Even stranger, objects that in postmodern times appear antinomic (plastic cups in zoo cafeterias, plastic rocks in zoo exhibits) often share the same extended qualities of form and function: human mouths that frame the rim, human eyes that frame the image. In zoos, which are strictly modernist, simulated trees and rocks can be manufactured to resemble natural objects, except that the models themselves—the conceptual precursors—are never transcendent.

The dictionary definition of *simulate* is to assume the outward appearance of something; a *simulacrum* is then an image or representation of whatever has been simulated. Zoos are populated by individual animals, and an act of simulation can impinge on what it means to be an individual.[3] In biology, *individuals* are organisms: a deer in your front yard, your goldfish, or you, but in all cases specific living objects. The term *individual* in metaphysics has a different meaning, including such disparate entities as a West Indian manatee named Snooty, the Boston Red Sox, London, and the constellation Orion. Snooty lives at the South Florida Museum. When his keepers toss Snooty a head of lettuce their intention is to feed him specifically, not the ontological *class* of manatees—that is, manatees in the abstract. Snooty is therefore both an organism (biologically speaking) and an individual (ontologically speaking). The pool in which he swims is also an individual and thus subject to change: it gets dirty, requires periodic water changes, and someday might break into pieces. By being mutable, its ontological status differs from that of *classes* of pools, which are changeless.

An ontological individual can be a single entity or an amalgamation of parts. Snooty, being a composite of cells, qualifies as a compound object. An individual's parts need not be contiguous. The Boston Red Sox baseball team is an individual even if its players are scattered across the country. The United States didn't lose individuality by granting statehood to Alaska and Hawai'i, neither of which adjoins the other states. Individuals ordinarily have proper names: Snooty, Boston Red Sox, United States. Species also assume individuality when we name them. Hans the Alpine ibex is an individual, but so is *Capra ibex,* his species.

An ontological class can designate an unlimited number of objects or none at all. Members of a class share only their defining properties: Snooty is a manatee, a class he shares with other manatees, even extinct ones. Both individuals and classes can have subunits, but with a difference. Whereas an individual is part of a larger whole, belonging to a class involves only membership. A class, in other words, is not part of anything. Finally, classes do not have proper names: manatees, pools, baseball teams.

When Captain James Kirk of the Starship Enterprise shouts, *Beam me aboard, Scotty!* his molecules are about to be scattered into black space then miraculously reassembled. We can assume Kirk survives as the same individual each time. At least he *seems* the same. The molecules, after all, are his own. Now imagine a slightly different scene, one proposed by Douglas R. Hofstadter and Daniel C. Dennett in *The Mind's I.*[4] As a space explorer you find yourself stranded on Mars, your spaceship wrecked and only one way home. A teleporter aboard the wreckage maps your molecules and beams the pattern to Earth where a machine reassembles it instantaneously from identical—but different—molecules stored in a generic molecule bank. Your own are gone forever. Would you still be you? Certainly, you reply. Some of my body's molecules are discarded and replaced every day. Over the course of my life every molecule of every cell—and the cells too—will change. But how important is it for the replacements to occur naturally over time, for the new components to be produced and organized into cells within your own genome?

Suppose, as Hofstadter and Dennett further suggest, a teleporter of a different design leaves the original you intact and sends an exact replica back to Earth, a clone so complete it includes a copy of your own mind. Your colleagues open the door at the receiving end and see you standing there. You look back, thinking what you would expect yourself to think in these circumstances, experiencing familiar feelings, memories, sensations. But is the you they embrace really *you*?

Now try these real-life situations. The Ise shrine in Japan is a Shinto temple built originally in the seventh century A.D. (fig. 7). In a ritual enacted every twenty years, its keepers destroy and reconstruct it. To the Japanese the temple is thirteen hundred years old, although no single piece is ever older than twenty years.[5] Is the shrine in its multiple incarnations the same individual or an extended series of simulacrums, each of them a different individual?

The *Charles W. Morgan,* now berthed at Mystic Seaport Museum in Mystic, Connecticut, is referred to as the last of the wooden whaling ships. However, she no longer puts to sea in search of whales, and the try-pots on her decks have not rendered whale blubber for many decades. Her crew is a crew of ghosts. Actually, she's incapable of being sailed. Long ago the museum's curatorial staff arranged for her lower hull to be stuck permanently in sand to prevent the incursion of shipworms. Much of her planking has been replaced over the years, not to mention numerous ribs and other important items of infrastructure. Is the *Morgan* an individual?[6]

According to David L. Hull, rate of change affects ontological status:

> Just as all the traits that characterize an organism at one stage of its development can change without that organism ceasing to be the same organism, all the distributions of traits which characterize a population at one stage in its evolution can change without that population ceasing to be the same population, just so long as such changes are gradual.[7]

Although the Ise shrine has existed conceptually for centuries, each reconstruction represents a new individual and a simulacrum of the original. The *Morgan* and I have changed continuously since gaining shiphood and personhood, but our structural organizations remain the same. Even if I were to lose an arm I would still be me. The *Morgan* devoid of shipworms and with new ribs is still the *Morgan,* just as a captive rhino minus its horn and after being fed a vermifuge retains its original metaphysical status. We are all individuals, unlike the temple that is razed and rebuilt and whose function ceases abruptly every two decades when its latest simulacrum vanishes. And the space travelers? Kirk is still himself (i.e., the same individual). The reassembled Mars travelers are different individuals, simulacrums of you. As the original, you might take a chauvinistic approach and claim

Fig. 7 Ise shrine. (Unknown photographer)

greater "value," just as an original work of art in Western culture is worth more than a facsimile. However, in the case of the second space traveler the replica is identical in every way. As Eco writes of the *Mona Lisa,* "the difference between the original and the copy would have antiquarian value (just as in the world of rare books the more valuable copy between two copies of the same edition is the one signed by the author) but not semiotic value."[8]

Jean Baudrillard, perhaps our most daring postmodern theorist, partitions simulation into three forms that evolved alongside society's technical expertise (or, in Baudrillard's words, "parallel to the mutations of the law of value").[9] These he calls *counterfeit,* or "first-order" (occurring from the Renaissance to the industrial revolution); *production,* or "second-order" (encompassing the industrial era); and *simulacrums of simulation* "founded on information, the model, the cybernetic game—total operationality, hyperreality, aim of total control."[10]

During counterfeiting, copies are made of an original. Production involves the organized manufacture of items, and "simulation" is modulated by social codes.[11] To clarify these concepts I quote from a short story that ends my book *Candiru*. A rainforest exhibit in a zoo is a counterfeit reproduction of a rainforest (i.e., a Baudrillardean "first-order" simulation):

> Although the components were produced and assembled by experts, nobody is actually fooled into believing the result is genuine. What we have is a one-of-a-kind counterfeit rainforest. This is because the duplication of any large complex object is impossible. Despite superficial similarities, no rainforest exhibit can duplicate a rainforest, and no rainforest exhibit can be an exact copy of another.[12]

In production (a Baudrillardean "second-order" simulation),

> items roll off an assembly line, each an exact facsimile. The model on which they are based might be a binary code or a sculpted object from which a mold was taken. Where the original exists—and in what form—is irrelevant. It might not even look like the final product. What matters is that each unit matches every other unit as closely as possible.[13]

Baudrillard's "third-order" simulation is hyperreality, which I describe as

> a dimension beyond simulation where reality has been produced from models devoid of any tactile substance or origin. No one can recall why such a thing happened or whether it ought to matter. The real is predictable in advance, rendering it indistinguishable from its simulation. In the hyperreal, deviation from either the real or the imagined is itself a simulation.[14]

As Baudrillard emphasizes, the model always precedes its simulation,[15] making the hyperreal similar to Plato's conception of the simulacrum (i.e., an identical copy for which no original has ever existed).[16] Any attempt at hyperreality is therefore a "third-order" effort to substitute simulated reality in reality's place, in the zoo's case to subvert an image of Nature by substituting one of its own. But the zoo never gets beyond counterfeit and production because its sole intention is to mimic—not transcend—the perceived true and real natural order. In contrast, "third-order" simulation is neither true nor false, real nor illusory, and always beyond control (any law that might control it is limited by being a "second-order" simulacrum).

Within hyperreality's boundaries fit the obvious candidates: most elements of Disney, holography, robotics, computer-generated "virtual reality," computer games, and cinematic special effects. Oddly, certain works of literature that conform to Baudrillard's criteria of "third-order" simulation are never mentioned in this context,[17] maybe because many consider the hyperreal to have arrived "with cybernetic society, when models began to take precedence over things, and since models are signs, signs now begin to exercise their hegemony."[18]

In addition, we have "reality" TV shows broadcast in the coded language of sex and avarice where watcher and watched are reversed, and manic dreams of childhood boredom are acted out. All photographs and movies are "third-order" simulations (their reproduction is "second-order"). So is the use of statistics in situations where the model is taken for reality, including declarative statements about such things as the "average" or "typical" American family, what Americans think about this or that, and probability in all its guises. Events, as interpreted and delivered by the news media, qualify for hyperreality (e.g., the destruction of New York's Twin Towers, the assassinations of John F. Kennedy and Martin Luther King Jr., the 1968 Democratic National Convention), and so do places as we perceive them in postmodernism (e.g., Los Angeles, Cancún, Las Vegas, and the West Edmonton Mall).

A form of representation common to zoos is the *artifact*, defined by Keekok Lee as "any entity or object which does not exist in nature (without human intervention), but is created by humans, according to human designs to fulfill human purposes."[19] Among her examples are domesticated flora and fauna: "A black tulip or a chihuahua is no less a human creation than a statue or a chair."[20] A living assemblage can also be artifactual—perennial gardens, for example, and zoological gardens. Lee distinguishes degrees of artificiality. Synthetic rubber is artificial, yet it functions in the same way as natural rubber. A silk rose, in contrast, could never function as a living rose. Synthetic rubber is a "second-order" simulacrum, the rose (a counterfeit) is "first-order." The black tulip, a pure artifact, resembles nothing except itself, and no previous model of it has ever existed. Like the dinosaurs in the film *Jurassic Park*, it lives in hyperreality.

The zoo, we are told, *is* Nature. Queue up, pay up, step into the wild. This is the zoo as simulacrum, the deliberate confusion of signs so that semiotic reality and vague recollections of television programs merge seamlessly. Support the zoo and you support the Other, an interposition aimed at making "zoo" and "conservation" synonyms. The postmodern spectator looks around at the aviaries, aquariums, and moated "veldts" and sees a coded wilderness complete with souvenir stands disguised as tiki huts, and English-speaking parrots chained to perches. Other animals reside among simulated tree trunks and boulders manufactured to resemble static television images or photographs from a magazine. True to a postmodern ideal that values self-perpetuation and redundancy, zoo guidebooks depict images of exhibits that themselves are reminiscent of images.

A quality strived for in "naturalistic" zoo exhibits is simulation of nature using artifactual objects (fig. 8). How logical is this? Not very. Nelson Goodman isolated what he terms "seven strictures on similarity." One of these applies directly to zoo displays: *resemblance is neither necessary nor sufficient for representation.* No picture (i.e., iconic representation) can be either more or less realistic or naturalistic than another. Such images—icons of a world unrevealed to us—no more resemble "real" Nature than paintings of the same objects. Furthermore, Goodman writes, "one dime is not a picture of another, a girl is not a representation of her twin sister, one printing of a word is not a picture of another printing of it from the same type, and two photographs of the same scene, even from the same negative, are not pictures of each other."[21] Surely "realism" can be separated along a continuum of the starkly real to the blatantly fantastic, but this notion is also false: "For pictures of goblins and unicorns are quite easily graded as more or less realistic or naturalistic or fantastic, though this cannot depend upon degree of resemblance to goblins and unicorns."[22]

A heavily planted and decorated exhibit of arrow-poison frogs is not more realistic than the same frogs displayed in a bare space.[23] The first is perhaps more pleasing esthetically, but supports no higher connection with Nature. As a representation it simply expresses someone's notion of what an arrow-poison frog habitat might look like. Is the educational value any greater? That depends on the message. If the text emphasizes how indigenous people use mucus from the frogs to make arrow poisons, an otherwise empty container will do supported by information about the process. If the emphasis is on camouflage, cryptic coloration, warning coloration, or something similar, then décor might be an asset.

Fig. 8 "Naturalistic" zoo exhibit featuring artificial rockwork and a man-made waterfall. Orlando, Florida. (Stephen Spotte)

We shall see in later chapters that zoo exhibits lack narrative and motion, and that an object must differ from our perceived reality to qualify as art. The zoo exhibit seems strange, not having been fragmented and reassembled in film-time. A novel invites mutual psychic invasion by storyteller and reader, and a film's *diegesis* (its narrative, or denotative content) remains unrealized by dispassionate observation. Fiction, whether read or observed, engages the participant, but the spectacle promotes only voyeurism. I mean this not in a prurient sense, but to indicate standing apart, observing without participating, maintaining hesitant space between object and eye. No other experience is possible because the scene is neither organized nor orchestrated. Although the décor might be manufactured to resemble natural objects and skillfully arranged, the focus is still the animal on display, and the animal itself is never treated as anything but real.

Zoos, in other words, attempt to display animals *as themselves*, hoping to redirect the spectator's gaze at a mental image of Nature as real and immediate. But because zoo animals are only semiotically real, the zoo's own gaze inevitably turns inward. Naming specimens in the collection or even giving them ascension numbers bestows individuality and distances them still further from Nature. Individuals in the Other are, by definition, unknown to us and therefore nameless.

Just like the exhibit décor around them, zoo animals are artifacts too. Eugenio Donato quotes Edward W. Said as saying, "What is given on the page and in the museum case is a truncated exaggeration . . . whose purpose is to exhibit a relationship between the science (or scientist) and the object, not one between the object and nature."[24] Museums and zoos are both repositories of many different kinds of artifacts—collections of semiotic signs—and any claim of homogeneity is necessarily spurious:

> The set of objects the *Museum* displays is sustained by the fiction that they somehow constitute a coherent representational universe. The fiction is that a repeated metonymic displacement of fragment for totality, object to label, series of objects to series of labels, can still produce a representation which is somehow adequate to a nonlinguistic universe. Such a fiction is the result of an uncritical belief in the notion that ordering and classifying, that is to say, the spatial juxtaposition of fragments, can produce a representational understanding of the world. Should the fiction disappear, there is nothing left of the *Museum* but "bric-a-brac," a heap of meaningless and valueless fragments of objects which are incapable of substituting themselves either metonymically for the original objects or metaphorically for their representations.[25]

A zoo animal, at most, is a semiotically real object and part of an exhibit that itself is a simulation. Metaphor shadows the notion that a captive specimen somehow stands in for its wild conspecifics, that preserving a species in zoos extends metaphorically to conserving wild specimens "out there" in the deserts or oceans. Seen from this perspective captive animals are indeed "bric-a-brac," isolated fragments whether of whole individuals or frozen bits of genomic matter. The flaw is in thinking that reality can be reconstituted through representation. A captive Alpine ibex *by itself* tells us as little about goats as the *Mona Lisa* reveals about women. Donato is correct when he writes that "representation within the confines of the museum is intrinsically impossible."[26] He further states:

> The museum can only display objects metonymically at least twice removed from that which they are originally supposed to represent or signify. The objects displayed as a series are of necessity only part of the

totality to which they originally belonged. Spatially and temporally detached from their origin and function, they signify only by arbitrary and derived associations. The series in which the individual pieces and fragments are displayed is also arbitrary and incapable of investing the particular object with anything but irrelevant fabulations.[27]

Anthony Vidler sees the fragment as important in two ways:

[1] As a reminder of the past once whole but now fractured and broken, as a demonstration of the implacable effects of time and the ravages of nature, it has taken on the connotations of nostalgia and melancholy, even of history itself. [2] As an incomplete piece of a potentially complete whole, it has pointed toward a possible world of harmony in the future, a utopia perhaps, that it both represents and constructs.[28]

Both might hold relevance to museums, but never zoos. Donato has demonstrated that museum objects can remain whole in their separation, and Vidler confirms that incompleteness imbues them with melancholy. The breakup of Nature, however, partitions wholes into smaller pieces that retain physiological homogeneity but devolve ineluctably to ecological ruin. An animal taken from a wild population, which in turn is part of an ecosystem, sees its metaphysical status reduced to that of a fragment, or the equivalent of a zoo-born conspecific. As Donato explains, such objects detached in space and time from their origin and function are capable of signifying as only artifacts can. As to Vidler's second signification, any thought of ultimate harmony through reassemblage—and thus utopia—is certainly shattered.

In museums of human or natural history—and in zoos where evolutionary theory is tacitly acknowledged in the arrangement of things—time and space shrink with alarming speed. Robert Harbison writes, "In museums . . . centuries are rooms and nothing but rooms."[29] Similarly, an entire phylogenetic class of vertebrates can be squeezed into a single building called a "reptile house," the possibility of perhaps thirty-five taxonomic orders into the "insect house." We build aquariums near abandoned urban seaports in an effort to "revitalize" them and restore a sense of nostalgia. Why should this be? Harbison tells us: "We visit the docks in London but not in Rotterdam, because commerce is romantic only when it has vanished."[30] We archive and display old furniture and marvel at it. After seeing such a collection in a European museum Harbison muses, "Looking

Fig. 9 Père David's deer, a living artifact. (Paul Massicot)

at a bed is not a transcendent act; it provokes low and irrelevant reflections: Would it be comfortable?"[31] What if Harbison's museum bed had come from a thrift shop? How could we know, and would it have mattered?

> The act of museumifying takes an object out of use and immobilizes it in a secluded atticlike environment among nothing but more objects, another space made up of pieces. If a museum is first a place of things, its two extremes are a graveyard and a department store, things entombed or up for sale, and its life naturally ghost life, meant for those who are more comfortable with ghosts, frightened by waking life but not by the past.[32]

Like museums, zoos collect objects after the fashion of modernism. Père David's deer (fig. 9) is an artifact according to Lee's definition. It exists today because of human intervention, and the only living specimens are in captivity or nature reserves.[33] When reproduction is controlled by "stud books" and range restricted to boundaries set by human beings, the surviving specimens are *de facto* human creations, domesticated animals in all but name. The American bison deserves such status, and so does the black-footed ferret,

Przewalski's horse, and California condor. N. Katherine Hayles argues for extending this designation to Yosemite National Park and other "wilderness" areas that survive only because of human intervention.[34]

By becoming artifacts these "rescued" species and others like them cease to be *natural kinds,* entities capable of evolving or even existing as independent agents. According to Lee, "Any organism which has been deprived of its reproductive powers and success may be said to be ontologically different from other naturally-occurring organisms and, therefore, lack independence and autonomy in a radical sense."[35] And as molecular biology intrudes relentlessly on steps taken to preserve endangered animals, every technological advance drives successive generations further toward the artifactual. Lee remains convinced that in degree of artifactuality transgenic organisms are on a level with plastic toys. Living versus nonliving is not the issue, but rather that both are composed of "existing natural kinds" and thus derivatives. Their *tele* is not their own, but acquired from their human creators.[36] The consequences could be the oblique denial of life if, as Benjamin claims, "Living substance conquers the frenzy of destruction only in the ecstasy of procreation."[37]

3

> One arrives in the garden again, at nursery time, when the gentle animals are fed and in all the world there are only toys.
> —William H. Gass, "Malcolm Lowry"

Zoos in postmodernism are trapped in their Baudrillardean "first-order" reality of counterfeit rocks and plants, of "second-order" captive-born animals that despite being alive are only semiotically real. Compare this situation to the rarefied space of hyperrealism where no detail is too small, no fantasy unattainable. What lifts the spectator into a realm of heightened illusion are the sensory accoutrements of postmodern technology and a vision that offers representations of Nature unattainable by any zoo. Are they "realistic?" Who can say? Reality and realism are not the same. Stated differently, how can live theater possibly mimic what "Nature" according to the cinema has already defined? Eco tells us bluntly, "Disneyland is more hyperrealistic than the wax museum, precisely because the latter still tries to make us believe that what we are seeing reproduces reality absolutely, whereas Disneyland makes it clear that within its magic enclosure it is fantasy that is absolutely reproduced."[1]

In Eco's opinion, hyperrealism can be extended to Nature strained through a zoo's message, if delivered artfully arranged and labeled as endangered species and ecology. I doubt it. Despite society's accelerating distance from Nature, the display is still too fleshy and immediate, the presentation too meretricious. While hearing the words, Eco fails to notice the messenger's shabby attire. Had he looked closely he would have seen how poorly Nature had been simulated, how incomplete the counterfeiting. At a marine park near San Francisco he watches as children are introduced to tiger cubs on leashes, where simulated Nature, like a Rousseau painting, is all color and light and noble animals cavorting in Arcadian gardens of timeless beauty. Here the prey might rejoice as predators strike them down. Nonetheless, even in a world of sodality you have to eat. Fortunately, the food offered to the carnivores in these putative Edens is already dead and thus unburdened by societal guilt. Still, eating, like defecation and sickness, is sufficiently modernist to keep hyperreality at bay. For transcendence we must look to Disney's mechanical crocodile.

The theme park is hyperreal, but not the zoo, and live animals when present are not defining factors even if they constitute the theme. As Michael Sorkin writes, the theme park is "the place that embodies it all, the ageographia,[2] the surveillance and control, the simulations without end."[3] What really matters is the happy face—joy and comfort without urban blight, poverty, crime, drugs, pain, dirt, and litter. In the effort to exclude reality, theme parks deliver the hands-off asepsis of TV and the movies, adding only a third spatial dimension.

Postmodern vision has also combined entertainment with shopping. At the apex of this frightening spectacle stands the West Edmonton Mall, covering 5.2 million square feet. There kitsch rules. Margaret Crawford describes the scene:

> Inside, the mall presents a dizzying spectacle of attractions and diversions: a replica of Columbus's *Santa Maria* floats in an artificial lagoon, where real submarines move through an impossible seascape of imported coral and plastic seaweed inhabited by live penguins and electronically controlled rubber sharks; fiberglass columns crumble in simulated decay beneath a spanking new Victorian iron bridge; performing dolphins leap in front of Leather World and Kinney's Shoes; fake waves, real Siberian tigers, Ching-dynasty vases, and mechanical jazz bands are juxtaposed in an endless sequence of skylit courts. Mirrored columns and walls further fragment the scene, shattering the mall into a kaleidoscope of ultimately unreadable images. Confusion proliferates at every level; past and future collapse meaninglessly into the present; barriers between real and fake, near and far, dissolve as history, nature, technology, are indifferently processed by the mall's fantasy machine.[4]

Another merger of animal exhibition and hyperreality is the Radisson SAS Berlin Hotel, which features a cylindrical aquarium variously described as twenty-five meters or sixteen meters tall (fig. 10). Among the many postmodern touches are "cameras which are integrated into the aquarium [to] show the fish life inside . . . to guests at the bar on large digital screens."[5]

Zoo exhibits are presented in fixed, modernist space. Postmodern space expands through compression, a logical paradox. In virtual reality, illusion of the infinite replaces normal perspective in a device that can rest on the bridge of a human nose, and our eyes, having adapted, see nothing unusual. This is cyberspace, a location where

Fig. 10 Cylindrical aquarium at the Radisson SAS Berlin Hotel. Sergei Tchoban, nps tchoban voss architekten BDA, Berlin.

"space" is actually nonexistent. Thus when "Main Street USA" intersects New Orleans' Bourbon Street both retain an unexplainable dimension that somehow passes for "landscape." Call it advertising, that address of the brain in which spectacle displaces contemplation, where one mirror held up to another reveals a different image. In postmodern times we discard the first image and cherish the second. As Benjamin anticipated, it is not "what the moving red neon sign says—but the fiery pool reflecting it in the asphalt."[6] Considering how the mall distorts Nature beyond recognition, any reflection suits.

No permanent harm is done by adding some tigers beside a lake containing robotic sharks and live penguins.

If zoos can be equated with wax museums as possessing a low degree of hyperreality (or none at all), then the supposition that people experience an emotional change when visiting them needs to be demonstrated. Wax museums often try to frighten us, but zoos attempt to induce, at minimum, contemplation.

At the zoo there are no characters with whom we can identify. Unless animals have been personified *we see nothing of ourselves*. The idea of Nature rendered clean and unthreatening holds only comfort. Dangerous animals can now be approached, having assumed human ways and human values. What Eco calls "universal taming" is a tension between "a promise of uncontaminated Nature and a guarantee of negotiated tranquility."[7] Whales "speak" on command, and this act of "communication" with their human trainers is tacitly imbued with meaning (fig. 11). When we tame a wild animal it ceases to be wild, becoming both a semiotically real object and a simulacrum of its former wildness.

Every zoo animal qualifies for "universal taming," but so do Jane Goodall's chimpanzees and the lioness Elsa. Having been named and observed at close range, they no longer represent wildness. According to Bob Mullan and Garry Marvin, "once the animal becomes the focus of human attention, the notion of a 'real animal' makes no sense for all animals as perceived by humans are the result of human interpretations."[8] Distance alone—the Other—can endow wildness. However, animals outside human culture lack individuality (and hence personality), and we tend to notice only general properties of their species. Mullan and Marvin state: "Lions, tigers, elephants and giraffes are animals which live in the zoo and although television might attempt to convince us that they have a *real* life in the tropical rainforests or on the African savannah this is not a reality when visiting a zoo."[9]

Naming zoo animals emphasizes individuality, placing them on a par with pets.[10] To then claim, as zoos often do, that such animals are "ambassadors for their species" is disingenuous. A captive dolphin named Flipper—marginalized, objectified, displaced, and now an individual—represents itself, not *Tursiops truncatus*. The same, of course, applies to its offspring. Elsewhere I've written:

> Out of sight, out of mind, out of the natural milieu—and out of circulation. When zoo animals reproduce, the offspring are second-order simu-

Fig. 11 Whales "shaking hands," but what does it mean? Katase Nishihama coast near Enoshima Island, Japan. (Stephen Spotte)

lations. Change—that is, *natural* change—is out of the question. Assembly lines can be re-tooled when necessary, but in biology we call this evolution.[11]

When Eco decided to take a trip into hyperreality he naturally came to the United States and made a beeline for California. He noticed that at Disneyland illusion is openly admitted: "A real crocodile can be found in the zoo, and as a rule it is dozing or hiding, but Disneyland tells us that faked nature corresponds much more to our daydream demands."[12] Eco doesn't say this, but we could interpret his statement to mean that Disneyland is more like a film studio where activity is edited, compressed, and presented in cinematic time. In Eco's opinion, Disneyland's coded message is that technology has a greater reality than Nature.

Richard Schickel had earlier reached this conclusion and proposed why manufactured Nature seems so appealing. Put simply, it induces momentary fear without negative consequences. Schickel writes: "What the average, middle-class American wants and has always wanted of art and of the objects he mistakes for art, is the fake alligator [sic] that thrills but never threatens, that may be appreciated for the cleverness with which it approximates the real thing but that carries no psychological or poetic overtones."[13] Today's urban humans are nervous in the unmonitored outdoors. Lee writes: "We only feel at home when home is the world of humanized nature—in other words, only when the natural has become transformed into the artefactual."[14]

The concept of superimposing an artifact onto a background of greater reality has interesting parallels in literature and art. This book is a collage in terms of its structure. The text, which is my own writing, is interspersed with quotations and endnotes identifying the writings of others. Picasso's *Still Life with Chair Caning* (fig. 12) was the first collage introduced into "high art" in the modernist era. Picasso inserted among the objects in view (e.g., a lemon, a newspaper) a swatch of oilcloth manufactured with a caning pattern commonly used on chairs. The effect was metonymic, suggesting a whole chair. The fragment of oilcloth is a semiotic sign signifying the referent, or object (a chair).[15] The pattern on the oilcloth is a simulacrum for chair caning and also an artifact. The other objects in the painting are easily identified signs, and the finished work of art is itself an artifact. Picasso later produced paintings onto which he glued objects. Here the fragments so used are not simply signs but actual pieces of the referents.

Theme parks are collages too, pulsating in postmodern 3D. The boat on which spectators ride is an artifact, perhaps produced on an assembly line (i.e., a Baudrillardean "second-order" simulation). If built by hand to look like the *African Queen* it represents a "first-order" simulacrum, a counterfeit *African Queen*. The watercourse is probably manmade and thus artifactual, although the water is real. Being "attacked" by a robotic crocodile represents a hyperrealistic experience, a simulation of high order. Despite the crocodile's simu-

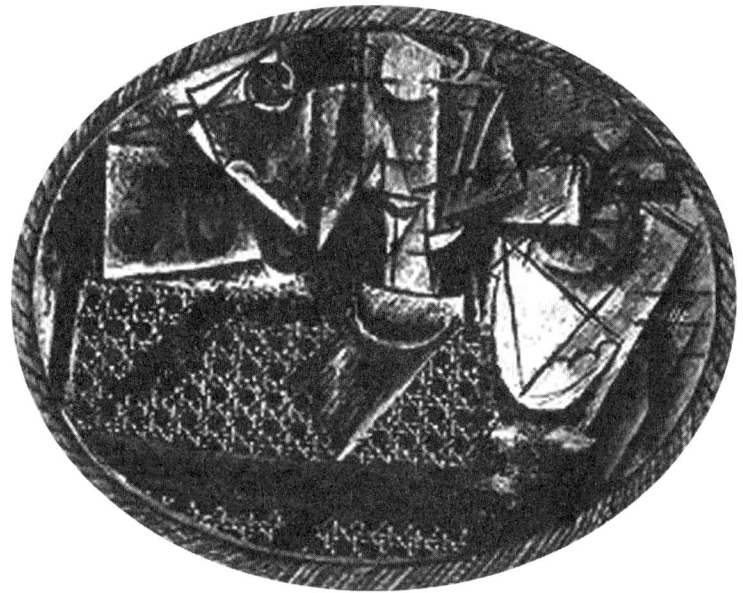

Fig. 12 *Still Life with Chair Caning.* (Pablo Picasso, 1912) © 2005 Estate of Pablo Picasso/Artists Rights Society (ARS), New York.

lated reptilian appearance, its robotic heritage makes it "third-order" instead of a simple forgery of a crocodile. Some robotic animals are realistic enough to signify established referents. The simulacrum of the white shark in the movie *Jaws* (named Bruce by the film crew) closely resembles white sharks filmed underwater (the referents for most of us). Robotic and computer-generated dinosaurs are another matter because no human being has ever seen a live one. This situation changed after *Jurassic Park,* and the current postmodern referents are cinematic images. We now expect our animals to behave as they do in the movies and at amusement parks, and when they don't they seem less real.

Zoo spectators know the animals are captive. By showing them in "natural" settings the exhibit designer tries to reproduce reality—the appearance of a habitat in which the animals on display might be found. The "Nature" offered is realistic without being real. As Eco says, at wax museums all things are signs that attempt to be real; at the live animal parks all appears real, but aspires to signs.

Of course, a form of self-induced virtual reality can be generated exclusively in the mind with only a printed page as stimulus. Julio

Cortázar propels zoo spectatorship into hyperreality in his stunning short story "Axolotl."[16] The unnamed protagonist begins visiting the aquarium at the Jardin des Plantes in Paris and becomes fascinated, then obsessed, by the axolotls on display. He stands for hours observing their slight movements. As his concentration intensifies, he begins to notice subtle details. He starts to *see* in the transcendent context of all observation: the toes and tail, the translucent tissue, and finally the eyes "lacking any life but looking, letting themselves be penetrated by my look, which seemed to travel past the golden level and lose itself in a diaphanous interior mystery."[17] He speaks to the axolotls, his mind resonating with imagined answers. All the while he denies their affinity with other animals while convincing himself that neither are they human.

Obsession becomes fear, and he welcomes the unobtrusive presence of the guard and the other spectators. Watching the axolotls drift in suspended immobility, he pictures himself devoured by their empty yellow eyes. He begins visiting the aquarium day and night and feels their suffering, their stiffness induced by the cramped quarters. Then suddenly: "No transition and no surprise, I saw my face against the glass, I saw it on the outside of the tank, I saw it on the other side of the glass. Then my face drew back and I understood."[18] He has become an axolotl too, but retains both his consciousness and human history. Like him, the other axolotls are secretly human, although all are mute and communication among them is impossible.[19] They can do nothing except think. And wait.

Representational thinking was an important breakthrough in Western thought after the Middle Ages. Semiotically, images (e.g., religious icons) and objects (religious simulacrums) could have realistic instead of metaphoric referents. To Eugene Hargrove this means that a painting of a lily could signify a lily, not just "lilies of the field."[20] Later, representational landscape painting imbued Western civilization with a sense of the esthetic that lasted three hundred years, well into late modernism and its highways with scenic overlooks. Scenes such as these, framed like paintings (fig. 13), are simulacrums.[21]

Zoo-goers treat captive animals as spectacles. In turn, zoos treat Nature as spectacle, and this is evident in how animals and habitats are presented. The diorama has never actually disappeared from zoo exhibition, and "naturalistic" displays are still designed to be viewed from specific vantage points, as if framed inside a landscape painting or highway overlook. Such a simulation—the exhibit as diorama—

Fig. 13 Rio de Janeiro, framed and viewed from an armchair. (C. E. Jeanneret/"Le Corbusier" 1961, 81) © 2005 Artists Rights Society (ARS), New York/ADAGP, Paris/FLC.

Eco calls the "crèche-ification" of reality.[22] To Italo Calvino the frame is an integral part of art and literature: "It allows the picture to exist, isolating it from the rest; but at the same time, it recalls—and somehow stands for—everything that remains out of the picture."[23] This questionable extension of representation might work in the arts, but not in zoo exhibition where the dichotomy is too striking. Who can imagine a simulated rainforest extending through the exit doors to a parking lot outside?

Hargrove makes an important point in stating, "Preservationist concern for the animals represented in zoological parks is a special variant of representational thinking in which the focus of attention is not on the particular but on the universal—that is, not on the individual animal but on the species."[24] And herein lies a problem for zoos. Try as they might to emphasize the preservation of *species,* zoo spectators care about *individuals.* Zoos talk about preserving genotypes in the interest of biodiversity. However, genes are semiotically opaque. What spectators pay to see is their phenotypic expressions. And of these, they respond most strongly to individuals, not classes. As shown later, humankind converses and speaks in metaphors, and our lives are defined by narrative. Each of us interacts on a daily basis with people, not humanity, and when we meet we tell each other stories. Individuals are people we know. Everyone else falls into the class of humanity, most of whom speak languages we can't understand and have customs that seem odd or terrifying. We shall never hear their stories, nor will they ever hear ours. When Pierrot, a character in *Aria de Capa,* Edna St. Vincent Millay's one-act play, claims to love humanity but hate people, most of us would think he meant the reverse.

It's therefore astonishing how a savvy zoo director like William Conway could find it strange that "We are touched with sadness at the plight of vanishing species but much more readily brought to tears by the difficulties of E. T., Dumbo, or Mickey Mouse."[25] From the argument so far it ought to be clear why Dumbo—an *individual* despite being entirely imaginary—elicits more concern in Western culture than the unknown, faceless *class* of real elephants inhabiting Africa and Asia. Moreover, both Dumbo and a live elephant at the zoo appear real to us despite being semiotic representations. Those vanishing elephants living in the Other are, by definition, inaccessible and unknowable. Without personal knowledge of them there can be little expectation of concern. This is the very reason why Save the Children Foundation shows us individual children in its TV ads and not hordes of children filmed at a distance. The poet Yehuda Amichai got it right:

> We do not wish to see the forest
> We wish to see the trees, the tree.
> The child, not the human race.[26]

Hargrove predicts that representational thinking might become unfashionable in the twenty-first century, just as representational painting (e.g., landscape painting) has long been passé. In the opinions of some, other forms of visual representation, such as Ansel Adams photographs and TV wildlife documentaries, are expected to follow.[27] Citing another source, Hargrove suggests that "narrative thinking" will supplant representational thinking in the coming years. This is a peculiar comment, considering that thinking and communicating in stories have always been characteristic of our species. Nonetheless, Hargrove continues,

> If narrative thinking displaces representational thinking, ultrarealistic replicas of natural habitats may not be enough to hold the attention of park visitors. The ultimate example of the narrative experience is the Disney theme park, which is little concerned with representational realism.[28]

He suggests, "In competing with Disney-style narrative, the display of living representatives of species may very well soon become the exception rather than the rule."[29] This idea folds into my argument that in their present configuration zoos are perhaps incapable of narrative, incapable, in fact, of reaching into hyperreality. Hargrove sees hope that a switch from representation to narrative will extend the influence of zoos into postmodernism, a doubtful prospect. Nonetheless, I agree with him that zoos are themselves endangered if they refuse to relinquish representation and take up narration in some postmodern form. Albert Borgmann describes the situation succinctly: "Accordingly we can now say that today the critical and crucial distinction for nature and humans is not between the natural and artificial but between the real and hyperreal."[30]

Zoos claim immunity from Disney-envy, aspiring to a higher purpose. This is both a false and a useless façade. In postmodern theme parks you can dine in "Paris" or "Madrid" or even "Bombay," except for one thing: no such referents are apparent. Not that it matters. As Paul Shepard writes, "Who cares about authenticity with respect to an imaginary origin?"[31] Similarly, zoos aspire to hyperreality with their replicas of imaginary undisturbed habitats. Were zoos truly content to stay modernist they might regress further and show representations of those degraded places predicted to be Earth's future: simulated mud and tire tracks and sawed stumps of trees, the whole devoid of wildlife.

In the simulation of unreconstituted nature—a world of nihilistic expectations—captive animals might even become superfluous. But unpleasantness makes for poor public relations. The result? "As fast as the relics of the past, whether old-growth forests or downtown Santa Fe, are demolished they are reincarnated in idealized form."[32] And if, as many believe, all wild lands have now been disturbed by humankind,[33] then a zoo exhibit presented as uncontaminated Nature based on images is like a photograph in a book: twice removed from the original, a reproduction of a reproduction, the model for which no longer exists.

American zoos have fervently adopted the "conservation" motif, in part to deflect the criticism of those who frown on keeping animals in captivity.[34] However, even their supporters will continue to hold Mickey Mouse in high esteem while setting mouse traps in their basements. An endangered mouse might garner at least token sympathy, until it moves in with you. Conway and colleagues fret about this seeming "paradox," but cultural modification is unlikely until narrative thinking takes the forefront. It appears that zoos have stood apart and sniped at their enemy without understanding him. From within Disney's hyperrealistic ecosystem Mickey exerts his alpha male status over zoos with impressive ease. Clearly, a rodent with human qualities deserves to be studied, not vilified.

The form of artifactuality apparent in zoo exhibits goes in and out of fashion. In the first chapter I described a sparsely decorated Alpine ibex display in which the animal and its surroundings were clearly divisible. Currently popular is the "immersion" exhibit, defined by Nigel Rothfels as "a place where both the animal and, increasingly, its human observer appear to be 'immersed' in a natural environment."[35] Although touted by zoo administrators as being original and a giant step forward in animal exhibition and husbandry, it is neither, as Jeffrey Hyson makes clear.[36] Such "naturalistic" exhibits have been around in principle and practice since the beginning of modernism. Furthermore, Hyson writes, "Zoo designs, like all works of landscape architecture, are clearly cultural constructions, yet the rhetoric of environmentalism may encourage the dangerous view that "immersion" exhibits actually *are* nature."[37] Zoos, at any rate, would like their spectators to believe this because it helps reinforce the idea of conservation being done both "here" and "out there."

One objective of "immersion" exhibits is evidently to elicit in spectators a particular "feeling," or Peircean Firstness (chapter 5), about Nature. Mullan and Marvin describe one display constructed in the 1980s to simulate an Asian rainforest:

The huge tree which dominates one of the areas is actually made out of steel tubing over which there is metal cloth which is itself covered by an epoxy resin textured and painted so carefully that most people would never guess that it is fake. But the vines which climb around it are real vines. Some vines however are *not,* and those which are provided for the gibbons to swing on are fibre-glass. The mist which envelops the tree tops is real mist but it is produced not by natural conditions, but by the sort of machine used in commercial citrus groves. The rockwork (except for the small pebbles) is artificial but it is a base on which real peat moss and algae grow. . . . The sound of the cooing of the forest dove is real but it was recorded in Thailand.[38]

These effects are less advanced technologically, less majestic and clever, than Daguerre's dioramas (chapter 7). The authors continue: "Although one certainly sees a jungle-like combination of flora and fauna, it is a secondary experience, or rather a primary experience of a replica jungle. It is an experience of what it might be like to be in a jungle as opposed to the experience of being in a jungle."[39]

But is a message about conservation best absorbed in such serenity, or is a certain amount of discomfort necessary? To *dissimulate,* as Baudrillard tells us, is to act as if we lack something we have. To *simulate* is to fake what we don't have. "Therefore, pretending, or dissimulating, leaves the principle of reality intact: the difference is always clear, it is simply masked, whereas simulation threatens the difference between the 'true' and the 'false,' the 'real' and the 'imaginary.'"[40] Both situations require the suspension of disbelief, but in different ways. The problem arises when zoos, which produce counterfeit ("first-order") simulations of Nature, attempt to force these exhibits into the hyperreal, clearly an impossible undertaking. The animals (so we are told) are contented because their surroundings now simulate the Other, that unattainable place where humankind has never been. How, then, is this possible?

In his essay "What is it Like to Be a Bat?" Thomas Nagel argues convincingly that we can never know what it's like to be other species.[41] Even if we could identify all their requirements we would still not know what it's like *to be* them. Consequently, zoo curators can only devise decorated enclosures in which the necessary spatial and social nuances, by definition, can never be known. Fooling captive animals in such settings is unlikely, as shown when Jabari, an adult male gorilla, escaped from the "Wilds of Africa" exhibit at the Dallas Zoo on March 18, 2004. Once in the public area he became aggressive and was shot and killed by police.[42] No gorilla, we could surmise, would ever want to leave Eden, a place where all the requirements of being a gorilla are provided. As the protagonist in Yann Martel's novel *Life of Pi* says, zoo animals "don't escape *to somewhere* but *from something.*"[43]

As a denotative entity, an "immersion" exhibit depicting a tropical rainforest has more in common with another zoo exhibit—even one displaying polar bears—than it has with a rainforest. The world can't be copied. In art, an attempt to represent reality involves only people duping other people who know they are being duped. Builders of "immersion" exhibits promote the notion that the animals inside are being duped along with the rest of us. People who believe this are only duping themselves.

Goodman refutes *trompe l'oeil* as representation, calling it "simple-minded" and pointing out that a man might be many things: atoms, cells, a fiddler, a friend, and a fool, but that none of these constitutes the object as it is. He writes, "If all the ways the object is, then none is *the* way the object is. I cannot copy all these at once; and the more nearly I succeeded, the less would the result be a realistic picture."[44]

Accordingly, "The plain fact is that a picture, to represent an object, must be a symbol for it, stand for it, refer to it; and that no degree of resemblance is sufficient to establish the requisite relationship of reference."[45] This is because representation's means of expression is denotation, which is independent of resemblance. The relationship is ultimately consubstantial, a photograph of a rainforest exhibit at the zoo being the same as the actual exhibit, neither of which differs from a photograph of a rainforest.

Semiotically, representation requires that the sign and its referent exist on equivalent planes of reality. In simulation, the sign subverts its referent: "Whereas representation attempts to absorb simulation by interpreting it as a false representation, simulation envelops the whole edifice of representation itself as a simulacrum."[46] The interpretation of the "immersion" exhibit thus has interesting metaphysical overtones for how Nature is affected. The power of images in postmodern culture is well documented (chapters 8 and 9), but what guarantees that the sign for Nature embodied in the "immersion" exhibit is real? In other words, who is to say that "immersion" exhibits, which presumably signify Nature, have any semiotic meaning at all? If (and here I follow Baudrillard's analogy of God) our idea of Nature can be shrunk to a paltry cluster of signs, then Nature itself is

Fig. 14 *Man's Fate*. (René Magritte, 1935) © 2005 C. Herscovici, Brussels/Artists Rights Society (ARS), New York.

nothing but a simulacrum. Lacking any referent, the Other becomes interchangeable not with reality but with itself.

Suppose that along with viewing an "immersion" exhibit of an African veldt we could also look through a window at the actual veldt. To enhance the illusion of reality we superimpose our exhibit onto the larger vista in a manner similar to one of Magritte's paintings (fig. 14). This picture within a picture, positioned against a window, is actually no more or less real than the scene it depicts. Similarly, our imaginary zoo exhibit is no less real than the veldt on the other side.

Simulation has rendered them both artifactual and incapable of transcendence.

Suppose further that the animals on the veldt and in the exhibit have all been bred in captivity. Releasing captive-bred specimens into the wild perpetuates artificiality[47] because the animals themselves are "second-order" simulations, having been conceived, born, and reared under controlled conditions. Their subsequent release into what for them is now an alien environment establishes still another level of simulation, particularly if subsequent observation, tagging, monitoring, or some other form of interference is called for. Only *as* the Other can Nature exist apart from us; compromise is impossible.

In the zoo as commodity (chapter 7), novelty stands as an independent quality. "It is," Benjamin writes, "the quintessence of false consciousness, whose indefatigable agent is fashion."[48] "Immersion" exhibits are not just unoriginal but historically so. They are, however, evidence of a pattern recycled through the cultural psyche as images of untouched Nature handed down by Romanticism. This is the zoo exhibit as eighteenth-century landscape painting, the difference being humankind's invitation to enter the scene. According to Benjamin, "The illusion of novelty is reflected, like one mirror in another, in the illusion of perpetual sameness."[49]

I wonder if spectators visit zoos to appreciate simulated Nature or just to see the animals. An appreciative response is less likely in Nature-based *trompe l'oeil* than it was in the early nineteenth-century dioramas where everything was artifactual but also nonliving. At the zoo, living animals roam through artifactual landscapes competing visually with objects created specifically for the purpose of drawing attention to themselves. Some spectators no doubt find this disconcerting. As when window shopping, they expect the merchandize to be emphasized, not the decorations. Modernist architect Victor Gruen, designer of the first shopping mall, saw a direct relationship. W. Jeffrey Hardwick, Gruen's biographer, writes:

> Gruen even fantasized about the ultimate in glamorous display methods, imagining future stores arranged like museum exhibitions. Spotlights would light single objects to accentuate their preciousness and patrons would focus on the individual items.[50]

Benjamin claimed that architectural space is experienced largely in a distracted state. Buildings, he wrote, are "appropriated" in two ways, by touch and by sight, and the first requires no contemplation.[51] During haptic appropriation we rush through doors and down corridors like rats in a familiar maze, mostly unaware of such details as a chipped brick on a wall, a burned-out light bulb overhead, or a potted palm in a corner. Distraction is the mental state in which the use of space becomes evident in our own movements, not contemplation. Proponents of "immersion" exhibits would have us believe that zoo spectators mimic the Chinese artist who, upon finishing his painting, stepped into it.

Many "immersion" exhibits are beautiful, but always inadequate representations of Nature, and always without narrative. Habitats and their biotic associations can never be simulated, nor can the disheveled beauty of a natural setting be reproduced even crudely. Benjamin warns that beauty is always problematic. Its relationship to the Other, he tells us while quoting Goethe, "remains true to its essential nature only when veiled."[52] The route is circuitous because the veil is what learned opinion decrees to be an acceptable reproduction of an object or image, thereby rendering the result always uncertain. By Paul Valéry's standards, "Beauty may require the servile imitation of what is indefinable in objects."[53]

Surely these definitions go beyond mere simulation, such as reproducing objects in molds and casts or copying them repeatedly from film negatives or digital components. Both activities falsify beauty by being "second-order" simulations. Without Nature's veil they can only be props, artifacts set among other artifacts that in sum are intended to serve as semiotic signs for Nature. But with the referent nowhere in sight they could just as easily be represented by plants clustered in the atrium of a shopping mall. A patch of green among the chrome and neon will never be the Other. And whether the plants are alive or made of plastic—that is, their ontological status—is irrelevant. A shopping mall, like a zoo exhibit, is an individual comprising parts that can be living, dead, or some of each, analogous to a one-armed man with a prosthesis.

4

> As the road to hell is paved with good intentions, so the road to confusion is paved with good metaphors.
> —Norman Macbeth, *Darwin Retried: An Appeal to Reason*

TZVETAN TODOROV TELLS US IN *THE POETICS OF PROSE*, "DEATH IS nothing but the impossibility of speaking."[1] Everything we do, say, write, or think—every facet of life as human beings—involves discourse. Zoo graphics are written discourse intended to serve a didactic function. Because writing is hard work a reluctance to suffer is often apparent, resulting in flaws of form and approach. The approach—static sentences placed before the spectator like text pages—shows contentment with being mired in the modernist tradition. Form can be addressed from the standpoints of narrative, textuality, and storytelling, the subjects here.

According to Steven Cohan and Linda M. Shires, "A *narrative* recounts a *story*, a series of events in temporal sequence."[2] The causality of a story is its *plot*. My contention is that a zoo exhibit is a *spectacle*, defined here in the context of literary theory as an event or series of events *without* a narrative because no story is apparent.[3] Parades, baseball games, concerts, carnivals, and circuses are other examples of spectacles. The recounting of the events observed at a spectacle can sometimes be transformed into narrative (e.g., the story of a baseball game can be recounted inning by inning). Following as a logical derivative, those who observe spectacles are *spectators*.

A zoo exhibit consists of a single, immutable event—its presence. It travels nowhere and seeks no change of venue. Because its principle function seems to be the static display of itself before a passing audience, any accompanying graphics offer the sole possibility of narrative. This is an opportunity seldom exploited. When zoo graphics disgorge facts instead of engaging the spectator in discourse their effectiveness as conduits of enlightenment is greatly restricted. Stories make human cultures cohesive by comprising the meanings through which we interpret where and who we are. They place us in our surroundings, give meaning to perceptions, and provide us comfort and identity. Movies, poems, novels, plays, songs, myths, anecdotes, commercials, sitcoms, videos, the evening news, and jokes are all stories transmitted in narrative. They speak to us, and we understand them. By omitting narrative and substituting description zoos place themselves outside the mainstream of meaning, threatening to confuse and disengage the spectator.

A narrative is not internal or private, but a public utterance, a fabric of signs woven in a way that allows them to be interpreted and understood by someone other than the author. A consistent weave is not required; it can be tight in places, loose in others. Several common textual devices serve to poke holes in narrative continuity and momentarily disrupt it, generating interpretants that apply imaginative force to the reader's experience. These devices—similes, metaphors, synecdoches, metonymies—add color and texture to narrative. Novelist Raymond Chandler, using a marvelous simile in *The Lady in the Lake,* describes a waiter as having "evil eyes and a face like a gnawed bone."[4] We never encounter such writing in zoo graphics, and zoos are the poorer for it. In talented hands, the definition of metaphor can itself be metaphorical. "Briefly," Goodman tells us, "a metaphor is an affair between a predicate with a past and an object that yields while protesting."[5]

Terry Eagleton makes the case that all language is metaphorical in the sense that words and phrases substitute for objects or ideas. He notes that metaphors save us from "the inconvenience of Swift's Laputans, who carry on their back a sack full of all the objects they might need in conversation, and simply hold up to each other as a way of talking."[6] George Lakoff and Mark Johnson wrote a book just about metaphors in everyday language, dividing them into groups (table 1).

According to Ferdinand de Saussure, meaning derives from differences between signifiers (Peircean signs), and these come in two forms: *syntagmatic* (when signs combine in a chain) and *paradigmatic* (when related signs can substitute for each other).[7] Syntagms and paradigms give meaning to signs by according them contextual positions within language structure. As seen in figure 15, these concepts can be displayed diagrammatically with syntagms on the abscissa (x axis) and paradigms on the ordinate (y axis). The syntagmatic axis comprises a series of words arranged in a correct combination (i.e., having proper syntax). A sentence is a syntagmatic expression, or syntagm, of words. Daniel Chandler tells us:

Table 1. Everyday metaphors of time and money.

You're *wasting* my time.
This gadget will *save* you hours.
I don't *have* the time to *give* you.
How do you *spend* your time these days?
That flat tire *cost* me an hour.
I've *invested* a lot of time in her.
I don't *have enough* time to *spare* for that.
You're *running out* of time.
You need to *budget* your time.
Put aside some time for ping pong.
Is that *worth your while?*
Do you *have* much time left?
He's living on *borrowed* time.
You don't *use* your time *profitably.*
I *lost* a lot of time when I got sick.
Thank you for your time.

Lakoff and Johnson 1980, 8.

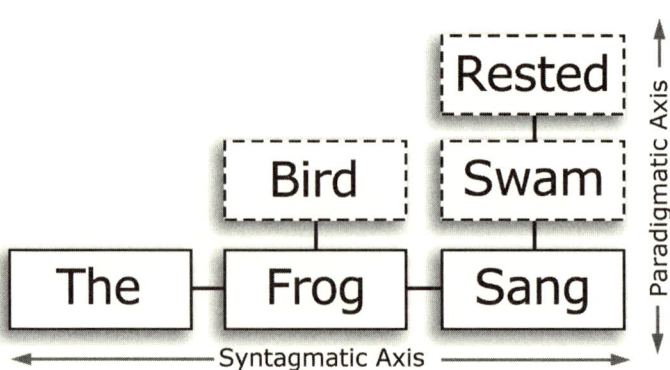

Fig. 15 Diagrammatic representation of the syntagmatic/paradigmatic relationship. (After D. Chandler, 2002, 80)

Syntagmatic relations are the various ways in which elements within the same text may be related to each other. Syntagms are created by the linking of signifiers [Peircean signs] from paradigm sets which are chosen on the basis of whether they are conventionally regarded as appropriate or may be required by some rule system (e.g. grammar).[8]

In a syntagm such as a sentence, one sign's position along a chain of signs determines its meaning, and meaning arises by the sign's asserting its *difference* from what precedes and follows it. "Whereas paradigms organize the vertical relations [ordinate] of similarity between one sign and others at the systemic level of language competence, syntagms organize the horizontal relations [abscissa] of contiguity between one sign and others at the discursive level of language performance."9

For example, Florida fishing laws dictate that only one snook (a saltwater game fish) can be kept during a day's fishing, and that a "keeper" snook must measure more than twenty-six inches in length but less than thirty-four inches. Such are the paradigms of snook fishing where I live. However, the syntagmatic situation is quite different. Suppose I catch a twenty-seven inch snook on my first cast of the morning. Instead of extending congratulations my fishing partner smirks and points out that although a snook in the hand beats a snook in the mangroves, if I release this one I might not catch another all day, and then any legal-size snook he catches will win our bet for "biggest keeper snook." My twenty-seven incher is obviously the biggest snook taken so far. Maybe neither of us will catch another. Do I keep it or release it and try for a bigger one? The paradigm states that I can keep a snook nearly thirty-four inches long. Considering the terms of our contest, a snook this size is more desirable than the snook in hand. Nonetheless, my smaller specimen has more value than might be surmised paradigmatically (i.e., the maximum length as described in Florida's laws) because of its syntagmatic position with respect to all the other snooks that might or might not be caught that day. And if I keep it I could still win the contest.

Specifically, paradigms are the groups, or sets, of associations that buffer signs with the rest of a language system. The conjugation of verbs is one example. The rules of baseball, axioms of geometry, laws of probability, Robert's Rules of Order, Florida's fishing laws, homonyms (words that sound the same), synonyms (words the mean the same), and antonyms (words that mean the opposite) are others. "Each of these paradigms locates a sign within the language system by structuring a relation of similarity, on the level of signifier [Peircean sign] or signified [Peircean interpretant], and this similarity helps to mark out the sign's identifiable difference from other signs."10 Cohan and Shires use as an example the homonyms *tale* and *tail*: "*Tale*, for instance, has the same phonetic signifier [Peircean sign] as *tail* but not the same signified [Peircean interpretant]; conversely, it has the same signified as *story* but not the same signifier. *Tale* is a distinct sign for this reason."11 They continue: "Phonetics, syntax, and semantics comprise major sets of paradigmatic relationships that classify groups of similar signs (nouns versus verbs, questions versus statements, animals versus fruits), and that differentiate one individual sign from another."[12]

A *phonetic paradigm* serves to distinguish *frog* from *dog* and *bog* by expressing both the similarity (—*og*) and differences (*fr*og, *d*og, *b*og). A *syntactic paradigm* places the sign (*frog*) in its proper location as a noun in the statement, *The frog is in the bog*. Finally, a *semantic paradigm* allows the sign *frog* to be identified among a host of such other signs as different animals, items, or phenomena (e.g., *dog, log, smog*), different amphibians (e.g., *salamander, toad*), and even nominative frogs (e.g., *Kermit, Froggy, Dan'l Webster*).

As Cohan and Shires point out, semantic paradigms isolate and distinguish meaning at the level of the interpretant, implying both recognition and understanding by the reader. The reader knows what *frog* as a sign represents well enough to separate it from other signs and conjure an interpretant: green amphibian, lives in ponds, occasionally spotted, jumps, ugly and fearsome to some, slippery, has long sticky tongue, eats bugs, progenitor of tadpoles, served as *cuisses de grenouille* (preferably with a cold chardonnay). Any of these interpretants can then be relayed, becoming once again a sign to someone (or something) else, triggering some sort of interpretant: *disgust* to some, *enemy* to a fly, the *evening's menu* to a French chef, *lunch* to a heron, *sex appeal* to another frog, *bringer of rain* to an Australian aborigine, *good luck* to a Japanese, *marker of environmental quality* to a scientist. These original interpretants, having been transformed paradigmatically into others, are now examples of *connotation*.

Note that on the ordinate of figure 15 any word can be substituted for any of the others. As Chandler says,

> In a given context, one member of the paradigm set is structurally replaceable with another. The choice of one excludes the choice of another. The use of one signifier (e.g. a particular word) rather than another from the same paradigm set (e.g. adjectives) shapes the preferred meaning of a text. Paradigmatic relations can thus be seen as "contrastive."[13]

Chandler reminds us that narrative might be the most obvious form of syntagmatic structure, and Cohan and Shires emphasize that syntagms "organize every aspect of discourse, from phrases of conversation to complete books."[14] Following the argument of these last authors, the statement *The frog is in the bog* is understandable, or has meaning, because of its paradigmatic structure. In addition, its meaning arises because of where *The frog* is situated syntagmatically. Placing these two words to the right of the verb alters the syntagmatic structure, producing a question: *Is the frog in the bog?* Word order is therefore important both paradigmatically (i.e., the syntactic paradigm identifying *frog* as a noun, the phonetic paradigm distinguishing it from *bog*) and syntagmatically (the placement of *frog* within a sequence of other signs).

The endless possibilities of combining words in English, and the equally daunting array of derivative meanings, comes about by the sequential arrangement of signs as in a sentence or paragraph. Although paradigms support this linear structure, the *value* of a sign (i.e., its relationship to other signs in the sequence) depends on its syntagmatic placement.

As creatures of language we read and write without being conscious of any of this, except perhaps in pedagogy. Unless a sign is unfamiliar we rarely stop to think about it, and only in extraordinary circumstances (e.g., while learning a foreign language) do we notice the interplay involving sign, referent, and interpretant. Instead, sentences are read and conversations assimilated for their global meaning. Interpretants arrive in a continuous stream. We divert them temporarily to memory, letting them spill into the next written sentence or spoken phrase where they asseverate new meaning, triggering subsequent signs and referents in a cascade of language. And all the while we barely notice, so immersed are we in the discourse. That is, until we encounter unfamiliar signs.

In chapter 1 I discussed Eco's hypothesis of cognitive type (CT), defined in simple terms as intuitive recognition or understanding based on previous knowledge or experience. The use of language is to transmit and receive information, but its evaluation involves comprehension and truth assessment.[15] For a statement to be understood it must contain semantic cohesion. *Semantics*—the meaning of language—is thus the purview of linguistics. However, for a statement to be true depends on the receiver's knowledge of the world, and this is the specialty of *pragmatics*.

As explained above, syntagms and paradigms confer meaning on signs by giving them contextual positions within language structure. Consider two statements: (1) *Goats generally have three legs.* (2) *Goats are three legs.* The first is correct semantically but in conflict with our world knowledge, which reminds us that most goats have four legs, not three. The second statement is incomprehensible semantically because the predicate *are legs* fails to coincide with how we remember and mentally erect the world. A goat, in other words, consists of more than its legs.

The facts presented in zoo graphics represent language transmitted in only one direction. Dialogue, which might foster questions that lead to understanding, is impossible. We can assume that most of the information available is correct semantically, but what about its pragmatic content? If we say or write, *Alpine ibexes have* flat/round/angular *horns and originate in the European Alps,* three different sentences are possible, differing only in how the horns are described. Regardless of choice (flat, round, or angular), all parts fit syntagmatically (i.e., in sum contain semantic comprehension), but paradigmatically only one is correct. Without previous knowledge of goats each choice has the same truth value. Zoo administrators might provide a species label or graphic with the correct answer (flat horns) and presume that still another fact has been archived in the spectator's world knowledge. A fact, however, is merely a contributing component to a CT, not a CT itself, which comprises many facts that in isolation have limited pragmatic value.[16]

Narratives make up both fictional and nonfictional accounts of events. *Fictional* accounts are considered to be "untrue," *nonfictional* accounts to present again (*re*-present) "reality"—that is, events as they actually happened. In practice the distinction is always fuzzy. In semiotics, for example, signs lacking material referents are fictive (e.g., a red traffic light, which means "stop"), but this distinction barely affects their semiotic reality.

Those who compose zoo graphics largely avoid fiction, perhaps believing it to be the dark side of truth, and, by extension, the "unreal" world. Literature, of course, is illusion.[17] However, all spoken and written means of describing reality fail the truth test and are themselves unreal because, as Wolfgang Iser reminds us, "the conveyer cannot be identical to what is conveyed."[18] Metaphors, for instance, can only be interpreted literally. This might be a frightening prospect were it not also true that much of science can only be explained metaphorically.

Storytelling—the recounting of a story in narrative—is clearly fictive. We know from the start that the sentences in a story do not denote the empirical world, although storytelling communicates with us nonetheless and instructs us about what we perceive. If Goodman is right, metaphorical and literal interpretation are equally informative.[19] Catherine Z. Elgin, a colleague of Goodman's, writes, "There are, to be sure, no rules for the interpretation of metaphors

[but] there are no rules for the interpretation of literal symbols, either."[20] Of the many voices we hear, the artistic voice is the brightest, most buoyant, and perhaps closest to the truth.

Even expository writing is considered a narrative form by some authorities. As Chandler points out, "*Exposition* relies on the conceptual structure of argument or description but it also has a narrative dimension."[21] His example is the type of expository writing found in academic journals in which sentence structure must be tight, the paragraphs linked cohesively, and the style transparent and seamless to prohibit any refraction of meaning. Banned are digression and discursiveness. The objective is to be accurate, clear, and concise.

In my opinion, calling expository writing *narrative* is too generous. As I restrict the contextual definition, *narrative* refers to a kind of discourse not yet applied in zoos, a thin ribbon of scientific description immersed and carried along in a broader fictive current of storytelling (table 2). The dual objective: engage the reading spectator while simultaneously transmitting a single, well modulated message. The method: mix scientific accuracy with literary prose form. The result: intriguing literary graphics that tell a story. Good writing expresses itself. The dictum that the mark of an artist is to make his brush strokes invisible is clearly false, as any admirer of Vincent van Gogh can attest. The same argument can be made for prose when fiction's opaque brush strokes are held up against transparent exposition.

Eagleton makes the point that works of literature are imitative "speech acts."[22] Their language, far from quiescent, is kinetic energy waiting to explode in the imagination. Their real function is performative, seldom descriptive, and always written to be heard even in the mind's silence. When animals are unable to speak to us directly, our own language must suffice.

We remember plot, not details, and when a text comprises nothing except details there is nothing to remember. We find this situation in zoo graphics. Iser quotes Eco, who argues that life is more like *Ulysses* than like *The Three Musketeers*. What Eco calls the "civilization" of the contemporary novel is its emphasis on future events and the unpredictable nature of its plot: "The event has not happened *before* the story; it happens *while* it is being told, and usually even the author does not know what will take place."[23] Thus a contemporary story sacrifices the mythic character of its predecessors (e.g., the trials of Hercules) with their predetermined outcomes, but gains humanity.

Table 2. Some elements characterizing narrative.

Unlike the "real" world, a narrative has a beginning and an end.
Narratives distill "reality" to discrete units of time.
Narratives create and define events.
What counts as an "event" is arbitrary.
The broadly predictable phases of narrative are *equilibrium-disruption-equilibrium.*
Narratives make the strange familiar through structure, predictability, and coherence.
Narrative (storytelling) is a fundamental human method of organizing experience into meaning.

D. Chandler 2002, 90-91, Metz 1974, 17, 20.

What happens to the hero could happen to us, and his feelings and reactions are what ours would be in similar circumstances.

Importantly, such humanity can be acquired without even being human. Bambi did it. The semiotic elements of a text form part of an assemblage of semantic features identifiable by their universality. Encased within the textual structure characters ride the plot to their destinations. Familiarity is what matters in narrative fiction. In a different context Eco writes, "By such features as 'human', 'animate', 'masculine', or 'adult' . . . one can distinguish a bishop from a hippopotamus, but not a hippopotamus from a rhinoceros."[24] However, by attributing these and other features to an entire host of characters it becomes easy to distinguish a deer from a rabbit—that is, Bambi from Thumper. The only prerequisite is making them "human."

Zoos take considerable pains to deny their animals an alternative existence, that of myth. The "deadly" serpent is presented as a valuable consumer of rodents, which is true but dull. Not every bat is rabid, although rabies can be as sexy as eating bugs in flight. Bat sonar is fascinating, but the bat in myth and legend makes a more humanistic story. Not even names are safe. Killer whales have probably become "orcas" in an effort to downplay their predation on equally popular dolphins and seals. As Baudrillard says, "It is this *fabulous* character, the mythical energy of an event or of a narrative, that today seems to be increasingly lost."[25] In its place is exposition. By ignoring myths about animals we risk rendering them benign—and boring.

5

> I suddenly realize I went all the way to the Bronx Zoo naked.
> Nobody noticed. Body is a word nobody notices. The word naked
> is naked. I cover it with the word cover. I'm still waiting.
> —Clarence Major, "Body Heat"

SOME CONTEND THAT WRITING CAN'T EXIST WITHOUT A READER. At minimum, the purpose of writing is to communicate. In organizing his message an accomplished author ordinarily uses what Eco calls "a series of codes that assign given contents to the expressions he uses."[1] For communication to occur these codes must be receptive to readers. A conscientious author brackets the anticipated knowledge of his audience and assembles his prose into a *model text*. Unless this is done the information is likely to stay locked inside his sentences and therefore meaningless. The recipient of a text properly designed and written for him is the *model reader*. A little later I take up the conceptual basis of the writer-reader interaction and deconstruct some actual zoo graphics.

According to Eco every text, regardless of its purpose and intended audience, must conform *explicitly* with a general model. Included in this model are a specific *linguistic code*,[2] a particular and consistent literary style, and predetermined limits on specialized jargon so as not to extend textual semiotics beyond the reach of the model reader. Zoo texts are not exempt from these requirements, yet anyone who visits several zoos in succession might find texts ranging from doggerel to scientific description. The model reader, if considered at all, is nowhere apparent, and the consequences are usually disastrous. *A text written with everyone in mind is written for no one.*

As Eco notes, any text is likely to risk interpretation against a fabric of codes different from those intended by the writer. However, if the reader's potential code system has been ignored then what results is a text prepared for an "average" reader. But what defines an "average" reader as compared with the "actual" reader? An "average" reader could be anyone, Eco hints, including "children, soap-opera addicts, doctors, law-abiding citizens, swingers, Presbyterians, farmers, middle-class women, scuba divers, effete snobs or any other imaginable sociopychological category . . . are in fact open to any possible 'aberrant' decoding."[3] Nonetheless, he cautions, don't expect too much from the results: anyone can interpret them however he chooses. A text this open is defined by Eco as *closed*.

Examples of closed texts are comic strips, John D. MacDonald's novels (e.g., his Travis McGee series), and newspaper advertisements. Their aim is to direct the reader down an established, well-marked trail oblivious to those who balk at going or might not understand the instructions. Their fate is to fall aside while the rest trudge on, the idea being that sufficient numbers will persevere to make the effort worthwhile.

MacDonald's novels found a huge audience, although it was one primed to receive them; in other words, the "ad" worked. Travis McGee is a beach bum and former pro football player who lives aboard a houseboat at slip F-18, Bahia Mar, Fort Lauderdale, Florida. He won the boat, named *Busted Flush,* in a card game. McGee earns an intermittent livelihood finding and recovering "lost" items for people who would rather not go to the police. For this service his fee is half the items' value. Once McGee commences sleuthing he encounters nasty people bent on blocking his path, is badly injured and nearly dies, kills his antagonist in a desperate fight, and makes love to a beautiful woman before losing her, all the while sputtering a crisp, surly philosophy at a pitch just below buffoonery. His body bleeding and battered, but with ethics intact, he crawls back to the *Flush* to recuperate, often minus the reward but with Boodles gin close at hand. This formula worked for twenty-one novels, the first called *The Deep Blue Good-by,* the last *The Lonely Silver Rain.* As *aficionados* know, a color is part of every title.

Another attribute of closed texts and similar escape entertainments (including popular films) is their redundancy. We prefer stories with accessible (i.e., redundant) plots populated by characters we come to know by their idiosyncrasies. Eco refers to these schemes as *iterative*. We dedicated readers of the Travis McGee series know that each novel is constructed from the same iterations and that variations are only in the details. We remember where McGee hides his money aboard the *Flush,* and we could find slip F-18 in the dark. Everything is familiar, and we like it this way.

Travis McGee is potentially available to anyone. So are the comic strip *Wizard of Id* and newspaper ads for breath deodorants. All can be poured through a sieve designed to separate the mainstream linguistic code of postmodern America from everything else. The result is disappointing only if you then expect MacDonald's plots to change abruptly, becoming like Danielle Steele's, or the Wiz to swear off sarcasm. As for breath deodorants, not everyone recognizes halitosis as a social liability, or cares, just as some readers prefer Steele or toss aside the comics. Significantly, any of these texts can be read as a love story, soap opera, comedy, or tragedy. When it comes to closed literature the model reader is, at best, a vague sociological composite of a *possible* reader perhaps identified from a marketing survey. Model readers of closed texts are difficult to define, but one thing they are not is "average." In practical terms, "average" readers don't exist, only "actual" readers or "no" readers.

In *open texts* each interpretation made along a chain of events is linked to the others. Events are *not* iterative. Every interpretation strikes against the others, reverberating in endless feedback if the clues and semiotic constructs can be found and deciphered along the way. These texts are amenable to deconstruction by their model readers, who in doing so wander along cresive corridors stopping whenever they wish to examine whatever pleases them. Here Eco offers James Joyce's *Ulysses* and *Finnigans Wake* as examples. Both novels are lengthy, dense, and difficult (*Finnigans Wake* might even be impossible).[4] However, they attract only readers prepared to assume the task of comprehension and interpretation despite inchoate plots that shift in and out of focus (or momentarily disappear) and language that explodes against the structural background like a fireworks display. Add to these pyrotechnics Greek mythology, opaque allusions to other literature and events of the day, and endless puns. And there's music: when you read Joyce you hear him too. A reader expecting naïvely to stumble across the path to meaning will surely never emerge from the maze. Here the model reader is embedded in the text, and quite possibly is Joyce himself.

Peirce's triad of sign, interpretant, and referent (chapter 1) is subsumed in another triad that he called Firstness, Secondness, and Thirdness. *Firstness,* which is associated with the sign, can be considered as a "feeling," a fleeting mood, an impression. Its manifestation occurs in a flash of color, an instant regret, a remembered taste. A feeling of Firstness recognized as relaxed, hazy contemplation often arises while listening to music. *Secondness,* associated with the referent, represents fact in the form of a symbol, icon, or index. *Thirdness,* thought of in context with the interpretant, serves to bring Firstness and Secondness together. This triad makes possible an unlimited semiotic progression: "As a First, then, the Sign . . . also acts as a Third, bringing the next Interpretant into a relationship with the Object."[5] To a hunter looking at the Alpine ibex exhibit described in chapter 1, the species label (semiotic sign) might also act to bring the next interpretant (a dream of ibex hunting) into play with the captive ibex on view (both a fact and an icon for other Alpine ibexes). This notion of endless semiotic progression is central to understanding literary theory, in particular the interaction of reader and text.

According to Iser, fiction communicates less meaning than effect by containing implicit constructions requiring the reader to imagine linkages of signs with interpretants. True meaning and literature can never match exactly, making fiction indeterminate and thus connotative and separating it from description, which is purely denotative, explicit, and governed by more restrictive rules of expression. But literature is also something else. According to Eagleton:

> Literature may appear to be describing the world, and sometimes actually does so, but its real function is performative: it uses language within certain conventions in order to bring about certain effects in a reader. It achieves something *in* the saying: it is language as a kind of material practice in itself, discourse as social action.[6]

Language is symbolic: "Symbols enable us to perceive the given world because they do not embody any of the qualities or properties of the existing reality . . . it is their very *difference* that makes the empirical world accessible."[7] This extends to the imaginary world. Each work of fiction represents a specific reality, the expression of which is subsumed in known semiotic systems that render it real to us but in no way makes it empirical. The memory of events (or the memory of categorical events) is somehow contained as stable molecular or structural entities instantiated in long-term memory. Moreover, traces of memories can be consolidated or dissociated and recombined, and the result recalled and activated. Memory is thus labile,[8] and what we see in our mind's eye while reading a novel or watching a film is the acceptance of new situations into stored categorical events.

The language of fiction is familiar in its mimicking of ordinary speech.[9] With empiricism impossible, fiction's symbolism must be complex compared with speech and speech's significant gestures, pauses, inflections, and situational nuances of place and time, all serving to augment the spoken word with parallel semiotic processes.

When successful, fiction is sophisticated and subtle. Signs provide the reader with materials and instructions to produce the interpretants necessary for meaning.

Importantly, fiction resembles an event, presenting the reader with the illusion of participating in something real *and in real time*.[10] The sort of event I have in mind is not a parade, which might be seen in its entirety from high above, but a baseball game in which the previous moment has gone past, the next one has not yet occurred, and only the present exists. Reality itself is unimportant in a universe where it serves as fiction's alibi.

Similarly, literature presents the reader with an endless sequence of signs, each modified by the preceding one. Words assembled into sentences form signs, arousing images that like illuminated ghosts become interpretants. The fictive text is elusive, revealed in a transient semiotic stream that leaves behind only faint traces of itself. And also like other events the only reality apparent is in the moment. The sentence before your eyes quickly dissolves into the background, creating space for the next new context—novel interpretants arising from previous signs metamorphosed into Peircean Thirdness. These, in turn, lose their luster and fade away. Thus the whole text is never in play at once, but apparent only as a proleptic vine in which tendrils of memory seek out connections with the signs ahead. What remains afterward is a feeling, an impression, or a Peircean Firstness: "Aesthetic value, then, is like the wind—we know of its existence only through its effects."[11] And what we ordinarily remember are the characters, plot, and general chain of events. Memories are always linked with context, and recall "marks the limit to which the linguistic sign can be effective, for the words in the text can only denote a reference, and not its context; the connection with context is established by the retentive mind of the reader."[12]

Works of literature do not exist until read; they are complex semiotic systems constructed of significations requiring actualization by the reader.[13] In this sense a novel or short story consists of a bundle of unrealized signs awaiting signification. The interpretants and referents occur nowhere except in the reader's mind, and each is surely different from the author's. Clearly, literature must not be evaluated solely on its presumed merit—that is, in isolation—but in terms of the reader as well. In Iser's model, a text exists between two poles, that of the writer (artistic) and that of the reader (esthetic). Somewhere between is the work itself, virtual "as it cannot be reduced to the reality of the text or to the subjectivity of the reader, and it is from this virtuality that it derives its dynamism."[14] The message travels in both directions, becoming actualized in the virtual void. Meaning is dynamic, not static, and everyone's experience is unique. Iser tells us, "It is generally recognized that literary texts take on their reality by being read, and this in turn means that texts must already contain certain conditions of actualization that will allow their meaning to be assembled in the responsive mind of the recipient."[15]

Most spectators lack the knowledge and experience to "read" a zoo exhibit with total understanding. We can nonetheless draw parallels between the derivation of meaning in literature versus in zoo texts. The question arises whether the current practice of making zoo graphics entirely descriptive and impersonal is an effective technique for stimulating imagination. Zoo prose, like scientific prose, seeks to minimize interpretation by eliminating similes, metaphors, and other textual devices and focusing instead on facts. The reasoning has been that the factual approach, by insisting on a greater degree of accuracy, has superior educational qualities. Significantly, this hypothesis has never been tested. Graphics based solely on description and being purely referential must necessarily jettison textual quality and risk not activating an imaginative response (i.e., Peircean Firstness). Referential texts, by definition, lack esthetic qualities.[16] Perhaps just as important, they are often deficient in the substantive elements necessary for semiotic viability. In terms of the capacity to actualize, a string of discursive statements about feeding habits, range, mating behavior, and conservation status is little different from a page taken out of a dictionary.

The question then becomes a matter of objective. Which is more important, transmitting many facts while eliciting no emotion (description) or eliciting emotion (narrative) while transmitting fewer facts? Is the casual spectator likely to recall a fact about an animal's biology or retain a dim, unspecified feeling of Peircean Firstness about its situation? And which of these is more likely to result in personal involvement? One thing we know for certain: good stories are innately understandable and hugely popular. The same cannot be said about lists of facts.

The negative side—leaving too much to the reader—leads to *subjectivism* in which every spectator actualizes a text uniquely (Eco's open text). The danger here is in actualizing significations outside the intended framework and losing the message completely in a thicket of individual interpretations. This is possible because author and reader together construct the fictional world, and the author's perspective is always different from the reader's interpretation. The author's artistic achievement lies in forcing the reader to discard his own views in favor of new ones, thereby experiencing the world in a newly imagined way.

The fiction writer's *perspective*—that is, the author's view of the world—must include the reader too. Naturally, this world is new and outside the reader's experience, although not so far removed that imagining interpretants and referents becomes impossible. Were author and reader to share the same experiences in the same way there would be nothing original, and the novel placed between them would have no novelty.

The four principal perspectives in fiction are those expressed by the narrator, the characters, the plot, and the reader.[17] Each originates at a different location in the narrative's imagined reality and moves toward a common meeting point called by literary theorists the *textual meaning*, or the *esthetic object*. By following these trails the reader arrives eventually at a denouement, the textual equivalent of understanding. Iser writes: "Thus, the reader's role is prestructured by three basic components: the different perspectives represented in the text, the vantage point from which he joins them together, and the meeting place where they converge."[18] And through this process the reader has imagined himself in a new light.

Whether a zoo spectator can suspend disbelief to such an extent is questionable. Presumably, this must occur before any message can be accepted and then integrated into the spectator's vision of himself (e.g., that of an active participant). But in postmodernism the story is paramount, and little can be lost by trying.

The descriptive text is a means of controlling informational content, but its discursiveness shatters the semiotic process and attenuates meaning. Signs left unactualized implode without interpretation. Although such texts seem clear and straightforward to zoo personnel who write them, the presumed message might have little meaning to its intended readers. In descriptive texts the objective is understanding. It should come as no surprise that descriptive messages are mostly uninspiring: instructions for operating a microwave oven are no different qualitatively than stating an animal's range or feeding habits.

According to Iser, the reader, guided by the text, assembles the meaning:

A reality that has no existence of its own can only come into being by way of ideation [signification], and so the structure of the text sets off a sequence of mental images which lead to the text translating itself into the reader's consciousness. The actual content of these mental images will be colored by the reader's existing stock of experience, which acts as a referential background against which the unfamiliar can be conceived and processed.[19]

Before understanding and interpretation comes communication. Esthetic experiences, unlike those mundane events of daily life, can occur only because they have been communicated. The appreciation of art in any form lies within an inherent opposition, or conflict. Iser says, "The more explicit the text, the less involved he [the reader] will be, and, in passing, one might remark that this accounts in great measure for the feeling of anticlimax that accompanies so much of what is called 'light reading.'"[20]

In good literature a conflict takes place in the reader's mind. The reader is drawn into its shifting tides, forced to choose, and carried along in the flow of narrative to the denouement, now in the context of a resolution. In narratives that are *overdetermined,* signification leads in several directions resulting in a layered effect and suggesting more than one solution. Meaning splinters, charging the reader with the task of reassembling the disparate elements into a cohesive structure in keeping with his own experiences. In other words, the reader is engaged, and with engagement comes commitment.

The *repertoire* of a text refers to the conventions needed to establish fictional situations within a familiar context of culture and literary allusion that the reader can actualize. The information communicated from author to reader is always new because no text is capable of importing the same real-life experience described in its written narrative. This loss results in a gain, that of communicating something to the reader that is simultaneously original and within his experience. In the case of rhetorical, didactic, and propagandist literature, "Such communications are only truly meaningful if these values are being disputed in the real world of the reader, for they are an attempt to stabilize the system and protect it against the attacks resulting from its own weakness."[21]

Texts that ordinarily accompany zoo exhibits consist mainly of description, but the repertoire commonly includes propaganda in the form of statements about conservation. Fictional narratives that might alert readers to a range of opinions have seldom been attempted.

The accepted literary procedures for organizing the various elements of a text's repertoire and actualizing them are called *strategies,* and they operate at the intersection with the reader. Zoo graphics often dispense with strategies entirely. As discussed later, this is apparent when complex information is summarized in a sentence or two, sometimes in the form of universal statements. The result is loss of effect at the expense of content. Denotation (what the semiotic sign represents) replaces connotation (other associated signs), and textual organization is overt, eliminating any chance that emotional interpretants could take conscious form. To paraphrase Iser, when literary texts use facts they do so to stimulate the reader's imagination and thus prevent the work from being rejected out of boredom.

Together, repertoire and strategy envelope the reader and within their confines permit him to visualize the textual meaning. The reader's involvement—that is, the layering onto the reader's consciousness of a text's repertoires and strategies—is necessary for any communication to occur. Mere presentation of the text is inadequate. As an active participant the reader *creates* meaning where before were only possibilities. Moreover, reading a literary narrative is unlike any other experience. As Iser points out, when we observe an object we stand outside it; when we read we follow "a moving viewpoint which travels along *inside* that which it has to apprehend."[22] Texts exist, of course, in which interaction by the reader is discouraged. In such cases it has been assumed that textual meaning will be transferred and assimilated directly. The thoughts expressed in Mao's *Little Red Book* were presumed to insert themselves neatly among the folds of the brain like crows settling onto a furrowed field. Texts, Marshall McLuhan tells us, are enclosed spaces: "For writing is a visual enclosure of non-visual spaces and senses. It is, therefore, an abstraction of the visual from the ordinary sense interplay."[23]

What Iser calls the *wandering viewpoint* is the reader's mental presence, enabling subjective movement through a textual labyrinth of shifting perspectives. Although the whole text is "wired" in place and theoretically available, only small portions can ever be actualized. Any synthesis is left to the reader alone, and any "facts" uncovered are his. Iser refers to this presence as "the point where memory and expectation converge."[24] A conventional zoo graphic, by being purely descriptive, limits the wandering viewpoint to denotation.

The humoral immune response, which tends to reject allogeneic tissues, has its metaphysical counterpart during reading when our minds are invaded by alien thoughts and apparitions. These presences also risk rejection, but if we accept and internalize them they become part of our consciousness. Each reading of a connotative text enriches a reader's mental background by occupying and illuminating the interstices of experience and carves new shortcuts to more rapid comprehension. We know this as learning.

The wandering viewpoint is an intellectual touchstone originating in our perceptions, principally vision. The scenes we encounter while reading are *imagistic,* producing mental representations as we imagine them, pictures in the mind's eye. Such processes are, of course, internalized; that is, they are *internal* representations accessible only through a mental process called *phenomenal experience.*[25] James Krasner has used cognitive theories of visual perception (chapter 10) to explain certain aspects of narrative theory. Krasner believes that writers portray scenes and situations that readers follow by constructing their own mental images in a streaming sequence, enabling the story, its characters, and the settings to appear realistic in the fictional landscape. Krasner calls this authorial construct the *narrative eye,*[26] and through its pupil the reader enters the author's mind.

Zoos seem compelled to give "truthful" accounts in the form of denotative texts, the apparent motive being the metaphorical subsumation of captive animals into distant ecosystems we barely understand and conservation strategies we have slim evidence will work. As explained in chapter 1, the natural world—the Other—is unavailable to the zoo spectator, and a zoo animal can only be semiotically real. By this thread of infinitesimal thinness hangs the zoo in postmodern times, which leads to a question: why should the so-called "truth" in a denotative message be any greater than a simple narrative "truth" rich with connotation? Neither is objective anyway. As Anthony Wilden points out, "Objectivity in communication is as imaginary as perpetual motion in mechanics."[27]

Signs, by definition, refer to something outside themselves. In denotative texts their signification is limited to the empirical object denoted (i.e., reference to the denoted object, or referent, and its indicated meaning are never detached). Thus restricted, the endless semiotic flow generated by connotation, as occurs in literature, is denied. Using connotation the reader restructures a panoply of textual signs into an array of interpretants limited only by her experiences and imagination. Zoo graphics are denotative when they list and describe exclusively. Description often contains jargon and scientific terminology, leaving meaning outside the experience of casual spectators. The result is loss of interest and emotional disengagement.

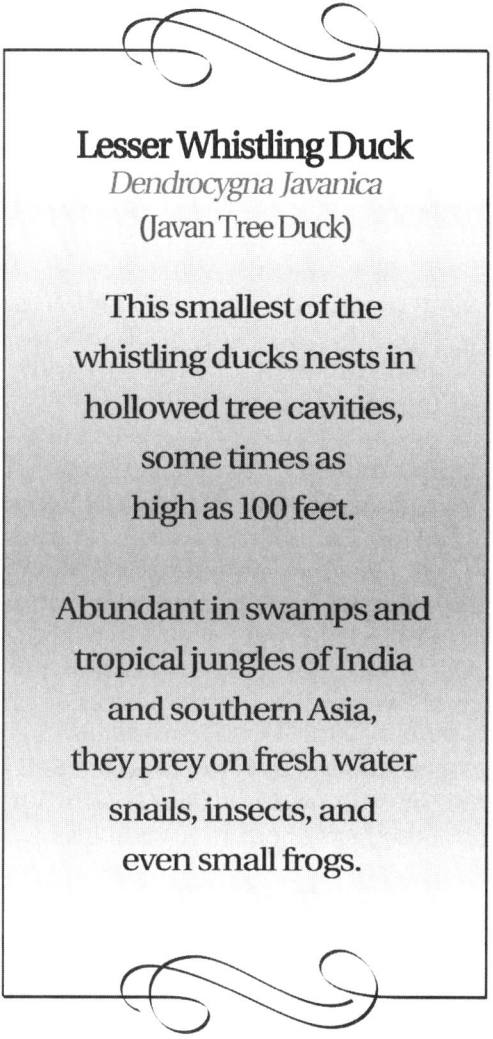

Fig. 16 A zoo graphic. (Stephen Spotte)

The text in figure 16 is reproduced from a graphic seen at a zoo in the southern U.S. Ignore the obvious flaws (the second sentence is backward and contains a plural subject) and note the content, which is purely descriptive. We are told (1) common name, (2) scientific name, (3) another (presumably secondary) common name, (4) size relative to the class of whistling ducks, (5) nesting habitat, (6) general habitat, (7) general range, and (8) food. A shaded range map nearby reinforces (7).

Some critics might classify the two sentences of text as narrative and claim they tell a brief story despite lacking all the commonly held attributes of literary narrative. A story has a beginning and an end, and this text has neither. Plot (as causality) is absent, and so is a convergence of perspective (the esthetic object). We could hardly hope to find overdetermination in two sentences unless the author is Franz Kafka. Repertoire is nowhere apparent, its place usurped by *negative allusion* in the form of potentially unfamiliar signs (e.g., swamp, jungle). Without a repertoire, organization and actualization—that is, a strategy—are impossible. And without any of these other features a wandering viewpoint (the reader's subjective movement through the text) is out of the question.

We can deconstruct this text on two levels. First, the absence of narrative renders the sentences bland, denotative, strictly factual, and minus any lasting impression. This is partly because none of the signs rises above any of the others, and because the paradigms are undistinguished. Furthermore, the content is barely heuristic, an assemblage of information. Nearly any of the paradigmatic entries could be replaced without altering the emotional impact: in a strictly denotative text the choices are limited to synonyms (e.g., "numerous" for "abundant," "wetland" for "swamp"). Second, because meaning and referent remain firmly joined throughout as perfect denotation requires, a coherent homogeneity is implied, linked with hints of a hidden *definition;* in other words, of the lesser whistling duck foregrounded against a few facts about its natural history. But if this text represents a sum of parts, what exactly *is* a lesser whistling duck? In truth, individuals (including species) defy definition,[28] exposing *Dendrocygna javanica* for what it is—a fragment of its own swampy habitat—but here at the zoo a disconnected metonymy, or a metaphor missing its referent.

Like most zoo texts this one assumes that its signs will induce a "cookie-cutter" series of interpretants in the minds of spectators. Such assumptions are nearly always false. "Jungle," for example, is defined in American dictionaries as "an impenetrable thicket or tangled mass of tropical vegetation." Its etymology (*jangala* in Sanskrit, *jangal* in Hindi) reveals several quite different meanings: forest, thicket, wasteland, country overgrown by long grass or weeds, wilderness, desert, arid land sparsely populated with trees. Given these many possibilities, what sort of "jungle" constitutes lesser whistling duck habitat? The words "smallest," "abundant," and "swamps" carry the similar risk of misinterpretation.

Without further description "Lesser Whistling Duck" is an empty sign, as is "The smallest of the whistling ducks" unless comparative sizes of the others are given. Because there are no true saltwater frogs the modifier "fresh water" before "snails" is redundant. "Javan Tree Duck" is another tautology (the common name "Lesser Whistling Duck" has already been presented), and "hollowed tree cavity" is still another. Nesting height is irrelevant without noting that many ducks nest on or near the ground.

The interplay between sign and meaning is a game from which the reader can disengage if a text becomes too obscure or boring. Nonetheless, a message should never be simple when the problem is complex; simplification often results in universal statements (chapter 11).

Before producing any graphics, a zoo needs to identify its model reader. Afterward, all writing and visual materials should be produced exclusively with the model reader in mind. I doubt that either is ever considered seriously. When reading fiction or watching a film the scenes roll by while the reader or spectator remains passive and stationary. At the zoo, animals are displayed passively and the spectators are moving. Thus the spectator controls the scenes, and this affects whether information (which ordinarily must be read) is assimilated. Ideas that work in the classroom, where the audience itself is held captive in a quasi-passive state, are unlikely to work during a visit to the zoo where such restraints are lacking.

To whom should a zoo direct its message? Specifically, what sociological composite constitutes the model reader? The answer must be known before either the form or content of any graphics can be effective. Obviously, a model reader who visits a zoo is someone interested enough to stop and read the information. Model readers can be thought of as intelligent, educated, and interested in gaining new knowledge through mental experiences. Usually they bring along some knowledge of natural history gained during previous learning experiences, but like anyone else they tend to ignore texts that bore them.

When the medium is the written word it assuredly is not the message too. Marshall McLuhan's dictum applies to electronic media, not to the ancient practice of writing. To extract a written message requires effort beyond the passive reception of film images and sound. In a printed graphic the message is the same whether screened onto a blue or a pink background or whether the font is Arial or New Times

Roman. These are unimportant details. What matters is *how* the message is delivered; that is, its form and content. My contention is that form should incorporate narrative and story, that content should be based exclusively on natural history, not science, and aimed at a targeted audience. Leave biology instruction to the universities.

So far as content goes, a zoo's graphics can, in principle, deal with science, natural history, or conservation. The temptation is to include all these, but hermeneutics and efficacy of the message might be better served if just one is selected. How to choose? A zoo whose function is to showcase the scientific results of its parent institution might emphasize science. Relatively few such zoos exist, and the scientific achievements of most of the rest are notably weak in theory (see chapter 11). American zoos promote themselves as participants in worldwide conservation efforts, but this is mostly posturing. That leaves natural history.

As to form, zoo administrators should ask whether they seek in their spectators an alteration of consciousness or see them simply as recipients of an information transfer. If the former, graphics would be more effective as stories than as description overburdened with forgettable information. The few attempts to date have been dismal. Despite incorporating a narrative of sorts, an example of the doggerel (fig. 17) produced by one zoo would have us think of its animals as white, politically correct, middle-class crib toys.

To engage the reader a zoo's message is best delivered as a story, and what facts are absolutely necessary can be embedded inside the narrative. Purely descriptive, denotative writing ought to be abandoned. "Fablephobia" is an ungrounded fear, considering that our lives are bathed in storytelling. I advise using plot, repertoire, strategy, and textual devices, and ignoring insincere lamentations like those of Bouvard and Pécuchet, Flaubert's confused bourgeois protagonists: "'The sun sets, the weather is overcast, winter approaches,' incorrect expressions which might make one think of personal entities when it is only a question of very simple events!"[29] I see nothing improper in asking what the weather is *doing* or in writing it. There are no "incorrect expressions," only a fear of dipping a rationalist toe into the unfamiliar waters of storytelling. Of course, most metaphors and similes are false in a literal sense. A charging elephant is no more an engine of destruction than a cheetah runs like the wind. Their use, as Elgin urges, "affords epistemic access to novel affinities both within and between domains."[30] They break through tired conceptual walls and open the world to strange delights.

Too often the writing of zoo texts is done by curators, scientists,

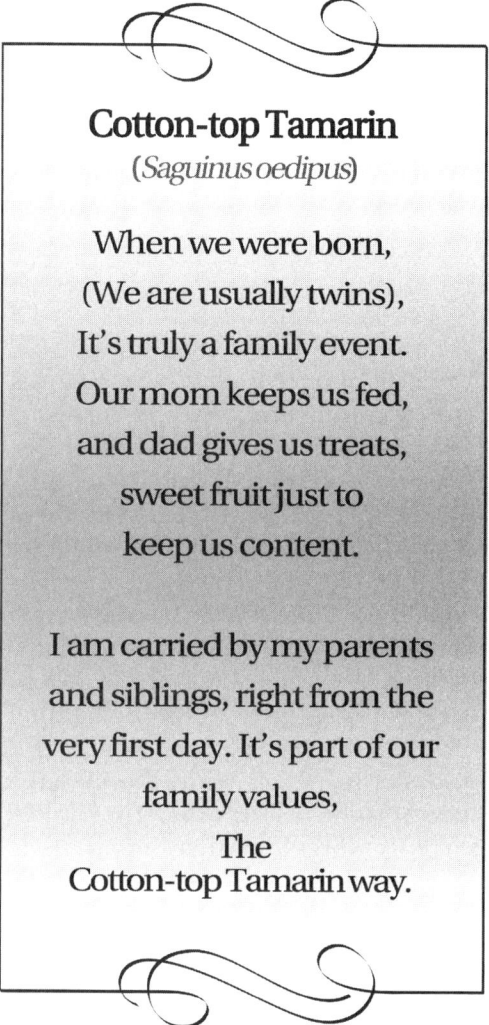

Fig. 17 Another zoo graphic. (Stephen Spotte)

veterinarians, educators—even committees—without regard to the targeted audience. Open texts, like strictly denotative texts, are an unappealing template for zoo graphics. Closed iterative texts offer several possibilities ranging from original fiction to fables and fairy tales. Before producing any graphics a zoo's administration must (1) distill its message to a few precious drops in a heroic attempt at clarity;

Table 3. George Orwell's six rules for writing.

1) Never use a metaphor, simile, or other figure of speech which you are used to seeing in print.
2) Never use a long word where a short one will do.
3) If it is possible to cut out a word, always cut it out.
4) Never use the passive where you can use the active.
5) Never use a foreign phrase, a scientific word, or a jargon word if you can think of an everyday English equivalent.
6) Break any of these rules sooner than say anything outright barbarous.

Wilden 1987, 260.

(2) explicitly define its model reader's boundaries of knowledge, interest, and ennui by conducting spectator surveys designed to yield unbiased results; (3) design the graphics for the model reader and no one else; and (4) use a professional writer to tell the stories, making certain that Faulknerian prose is avoided and George Orwell's six rules are followed (table 3). I would add to Orwell's list, *Never explain a story, but construct the narrative so that explanation becomes a seamless part of the process.* Good writing is identifiable by what it assumes.

In true postmodern tradition think of the narrative as being primary and information secondary. A story in need of explanation becomes indistinguishable from news. As Benjamin remarks, "Actually, it is half the art of storytelling to keep a story free from explanation as one reproduces it."[31] It is critical to elevate narrative above the mundane transmission of information, modernism's stifling legacy to zoos, while not forgetting that the problem of translating a scientific text into narrative is no different from the translation of one language into another. Both must place a priority on the artistic outcome, and both are next to impossible. To quote Benjamin's unyielding viewpoint: "Yet any translation which intends to perform a transmitting function cannot transmit anything but information—hence, something inessential."[32] Still, we can take comfort in Frank McConnell's reminder about *all* storytelling being didactic.[33] That we learn from stories is not in doubt. That we learn from compilations of forgettable facts is less certain. What matters ultimately are the questions asked about Nature, not the answers, which are derivative and always contingent. In an ideal zoo graphic the *story* is the message.

6

> I wanted to explain to him that what television and bad American movies had done was to make us doubt that others even existed except as a shadow play.
>
> —Barry Hannah, "Carriba"

Zoo texts are modernist by being discontinuous in an unfamiliar way. Postmodern spectators are accustomed to television's sequential cut, the jarring integration of commercials, news, sitcoms, and documentaries that embrace discontinuity and reassure us that reality is a pattern of unconnected images. As chain-smoking lady detective Mike Hoolihan says in Martin Amis' novel *Night Train*, "They want commercials every ten minutes or it never happened."[1] Even a TV commercial usually tells a story, but the zoo text resists. Moreover, text and graphics force us to read and interpret; TV reads to us while simultaneously presenting familiar moving images of our own species. Movement in our postmodern minds equates with narrative, paradigm and syntagm with flickering images in a darkened room.

The current unease with zoos goes deeper than text, graphics, and simulated objects. It extends to the animals themselves and their presentation in real time. Static exhibits have been passé since the invention of cinema. The placement of interactive exhibits in zoos is actually an attempt to introduce motion, and the construction of Imax theaters in close proximity to some zoos attests to a sense of incompleteness.[2] Motion condenses time, making it seem immediate. Still images, in contrast, are timeless. Motion is what makes cinema interesting, and motion is largely missing from zoological displays.

Zoo administrators have traditionally taken comfort in the unrealized—and undocumented—belief that spectators would prefer seeing the "real" animal to that animal's moving image. Such an attitude is dangerously anachronistic in the postmodern age when, in Guy Debord's famous phrase, "the image has become the final form of commodity reification."[3] The truth is, our lives are flush with images of all kinds, and those that move are especially welcome and attractive.⁴ Vidler writes,

> it is obvious that film has been the site of envy and even imitation for those more static arts concerned to produce effects or techniques of movement and space-time interpenetration. Painting, from Duchamp's *Nude Descending a Staircase;* literature, from Virginia Woolf's *Mrs. Dalloway;* poetry, from Marinetti's *Parole in litertà;* architecture, from Sant'Elia to Le Corbusier, have all sought to reproduce movement and the collapse of time in space; and montage, or its equivalent, has been a preoccupation in all the arts since its appearance, in primitive form, with rapid-sequence photography.⁵

Understanding a fiction film requires little imagination or effort. Add to this its matchless combination of action and visual impact and there can be little doubt about which medium of communication has influenced postmodern culture the most. Folded into the ensuing discussion is the documentary wildlife film. It would seem that this genre is especially pertinent in a book about zoos, but I disagree. The ontology of all images—wildlife films included—denies them representation of reality. As discussed before, what constitutes a "realistic" image of a unicorn is cultural, no more than an arbitrary place on a false scale of unreality.

Filmmaker Derek Bousé argues throughout his book *Wildlife Films* that what we consider to be documentary is actually "narrative entertainment."⁶ Wildlife films and cinematic fiction therefore share a common basis, and neither is more nor less selective than the other. For example, both forms are based on what I shall call a *closed-text cinematic format* (film's equivalent of literature's closed text). Both fragment and compress time, are scripted, use plot and narrative as cognitive glue, and rely on the same staging and camera techniques (e.g., *découpage,* close-ups, montage, fades, cuts, separation, slow motion, parallel action, cutaways, manipulations of *mise-en-scène*). Even *ellipsis* (the omission of action without including covering shots) and *eye-line match* are common in wildlife films. In an eye-line match the camera might first show a bull elk looking left before cutting to a female elk nearby. The female, if supposedly interested in the male, must be looking right, or in the expected direction to see the male. Implicit in all these techniques is the spectator's intuitive understanding. If we accept Béla Balázs' conceptual premise, films humanize Nature because selection of the landscapes filmed as backgrounds for our stories are products of culture.[7] It seems that film speaks to us, regardless of genre, and we seldom miss its meaning.

Our problem of understanding zoos in the postmodern world is marred by an irreversible addiction to motion. If we can attribute to zoo animals any actual elements of reality they would be silence and stillness. Film, in contrast, distills the intermittent flashes of motion characterizing the lives of most animals and presents them to us as "real." During a typical day African lions might rest or sleep twenty hours. This amount of time, if included proportionately in a one-hour TV film (fifty-two minutes with commercials), would occupy forty-two minutes.[8] Broadcasters and advertisers, of course, want only action. This is easily done through *montage*, the editing of one action sequence after another—first a lioness stalking prey, charging, making the kill, watching her cubs feed—all presented as if in real time. Actually, the scenes might have been filmed over weeks or months; alternatively, the "single" lioness might be a composite of several different animals. The finished product deludes its audience into believing that animals lead rich, eventful lives when in fact most of their time is spent inactively. But action is what we expect of our cinematic creatures. Editing out inactivity, shortening and splicing action sequences, adding breathless voice-over narration, dramatic music—all sharpen the illusion of speed, quicken and compress time; form *becomes* content. As Bousé writes:

> The idea is to present audiences with something recognizable, for which they already have conceptual categories; to be consistent with their previous viewing experiences, to fulfill not thwart their expectations, and to do all this by employing already familiar conventions of realism, not by trying to reproduce reality itself.[9]

The denoted natural world can never be realism *re*-presented, only fiction with a different label. Even place becomes fictive, a tropical rainforest on film looking nothing like it did on the set. In the film we see flowers open suddenly, rain drops magnified, the leaf litter turned over and illuminated, ants the size of sand grains filling the screen. But that's *us* pushing aside the branches to reveal some secretive animal with obtunded reflexes, not the camera, just as *we* gently crack a bedroom door and enter the cinematic lives of two lovers who also seem unaware of our presence. In closing these cognitive distances, and by participating, we banish any last traces "realism."

Wildlife films connote, suggestive of artistic bias and no different in this respect than cinematic fiction. From the standpoint of film theory I find nothing to distinguish them. As Christian Metz says, "Every film is a fiction film."[10] As proof, each tells a story.[11] Wildlife films are stories about animals told by humans, but most people are more interested in viewing their own species directly. Much of my focus is therefore on openly fictive films, which have hypostatized the postmodern consciousness until, as Geoffrey O'Brien writes, "What you thought was a movie theater turns out to be home."[12]

At the start of his excellent book *Narration in the Fiction Film*, David Bordwell reminds us that Aristotle, in his *Poetics*, divides imitation into three parts: *means* (a medium such as painting or language), *object* (some aspect of human action), and *mode* (how something is imitated). Bordwell interprets Aristotle's subsequent discussion on mode as distinguishing between telling and showing:

> *Diegetic* theories conceive of narration as consisting either literally or analogically of verbal activity: a telling. This may be either oral or written. . . . *Mimetic* theories conceive of narration as the presentation of a spectacle: a showing.[13]

Bordwell remarks that because the distinction between diegesis and mimesis is limited to modes of imitation, either can be applied to any medium (e.g., painting, literary fiction, film, theater).[14]

All forms of communication can incorporate "show and tell," although in spectacle one of them might be absent; if both are present the result is redundant. At a baseball game the announcer can say, *Casey at the bat*. However, if we happen to be at the game or watching it on TV we can see this for ourselves because Casey's name and number have been stitched onto the back of his jersey. The announcement and image together are tautological. Both are signs pointing to a referent (Casey at bat) and trigger a familiar mental image of Casey in his uniform (interpretant). If we are listening to the radio and unable to see Casey the announcement is not redundant, but neither is it both show and tell, it's only tell. Similarly, having a volunteer standing at the Alpine ibex exhibit and telling spectators, *This is Hans, an Alpine ibex* is redundant if the same information is available in written form nearby. The graphics will identify Hans as an individual distinct from other specimens of *Capra ibex*, just as Casey's jersey identifies Casey as an individual among all other specimens of *Homo sapiens* attending the ballgame.

Early film theory emphasized mimesis, probably in part because the first films were silent. Bordwell, citing V. I. Pudovkin's 1926 monograph *Film Technique*, explains:

> According to Pudovkin, the camera lens should represent the eyes of an implicit observer taking in the action. By framing the shot a certain way, and by concentrating on the most significant details of the action, the director compels the audience "to see as the attentive observer saw." The change of shot will then correspond to "the natural transference of attention of an imaginary observer."[15]

This notion of what Bordwell calls perspective from the standpoint of the *invisible observer* is analogous to perspectives traditionally used in static imaging like painting and photography, and also in the design of zoo exhibits. All take into account what the painter, sculptor, photographer, or exhibit designer envisions when standing behind an imaginary observer and looking over her shoulder. Of these, the zoo exhibit most closely fits Aristotle's concept of mimesis, that of a staged play (with animal actors) in which objects can be viewed within a range of fixed angles and presented in real time.[16]

Zoo exhibits also resemble theater stages in their use of backdrops and stage lighting. That the décor is devised and arranged largely to benefit the spectators is undeniable. It is "part of the ensemble of props needed for this theatre of illusion."[17] It is, in fact, theatrical furniture. As seen in figure 18, the stage is the horizontal part of the exhibit, its décor the carefully arranged boulders and wooden shelter. The moat separating the spectators has its theatrical counterpart in the orchestra pit. To extend these associations, what lovers of theater like best is the actors' presence—in this case live bears.[18] Certainly a zoo without animals on display would be incomplete.

Theater and zoos have in common the demeanor of subdued spectacle. Eco emphasizes that at "serious" cultural events the audience does not participate: "It sits and listens, or watches; in this sense a spectacle (or what was once a spectacle in the 'bad' sense) can become 'serious' when the public takes no active part but simply attends passively."[19] Eco reminds us that in ancient Greece the spectators at staged comedies spit fruit pits and taunted the actors. In postmodern times teasing animals or denigrating live actors is considered bad behavior.

Theater also shares with zoos the notion of dramatic place. As

Fig. 18 Modernist zoo exhibit circa 1900, which resembles a theater setting. (Unknown photographer) Denver Public Library, Western History Collection, X-27342.

André Bazin states, there can be no theater without architecture,[20] meaning the staging of contrast between the world outside and the world within the play, and because of this *locus dramaticus* there is décor: "It serves in greater or lesser degree to set the place apart, to specify. Whatever it is, the decor constitutes the walls of this three-sided box opening onto the auditorium, which we call the stage."[21] The area of the stage containing décor "is thus an area materially enclosed, limited, circumscribed, the only discoveries of which are those of our collusive imagination."[22] This space envelopes the stage and faces the audience. Baudelaire preferred looking at background paintings used in stage plays because, "Those things, so completely false, are for that very reason much closer to the truth."[23] Zoo exhibits are often constructed with décor used similarly, except the purpose is to draw nature inside, not keep it out. Desert reptiles are shown among sand and rocks, rainforest species underneath leafy canopies.

Bazin emphasizes that film rejects the idea of dramatic place as too confining and gives us a metaphor of space as cinematic protagonist. That a stage actress would deliver her lines while offstage is unthinkable, yet a movie actress disappearing off-screen is nothing extraordinary. In fact, we might leave her in Paris while the scene shifts to New York, unseen but not forgotten. Like badly trained actors, zoo animals often hide, much to the annoyance of spectators. *Where is it?* they ask. *Is anything in there?* If the star of a zoo display is naturally cryptic, a camera might be placed inside its den so that its activities can be tracked "outside" the exhibit. Filmed theater has never been successful, and filmed zoo theater fares no better.

The movie camera can go anywhere. Towering vistas and the recondite behavior of microorganisms are both accessible. Unlike plays, which require live human beings, film can get by on wildlife alone. A zoo exhibit, having been conceived and executed by humans, now excludes its creators, and the animals themselves—the actors—are alien and voiceless. Were it a window in a wall separating the spectator the result would be the same. The theatrical presentation of animals is based on a questionable premise that staging simulated Nature in real time can be both interesting and didactic.

In Pudovkin's model consecutive cuts mimic the gaze of an invisible observer, who perhaps is watching people meet on the street and shifting his glance from one to the other. Shortening the scenes could represent the observer's heightening excitement. Importantly, editing stresses continuity, "a witness who turns his attention from one detail to another; rooted to the spot, the witness remains on the same side of the axis of action or '180° line.'"[24]

There were other postulates. In one of these the invisible observer *became* the camera, which then metamorphosed into the narrator. To Pudovkin the lens and the eye of the director were indivisible, but others later saw the camera representing a disembodied storyteller expressing the director's viewpoint. "Thus the invisible-witness model became classical film theory's all-purpose answer to problems involving space, authorship, point of view, and narration."[25]

This model had flaws. One was ignoring stylized techniques in which such rigidity is impossible, another forgetting that any film contains some unrealistic camera angles.[26] Subsequent theories from this early period fared no better, and all ignored or minimized the spectator's participation, seeing him as a sort of sponge passively absorbing images as they diffused outward from the screen.[27] Either that or a programmed dummy waiting to be fooled, delighted, or saddened by the film director's clever cues. In fact, the spectator participates in a way similar to a model reader (chapter 5), although less imaginatively. Not surprisingly, a film director has greater control over our nonaffective experiences than a novelist might. As George W. Linden asks when comparing the novel with film, "And what is imagination except seeing as if present that which is not present?"[28]

Roland Barthes applied de Saussure's linguistic work on the semiotics of language to other cultural phenomena besides film, including professional wrestling, high fashion, and advertising.[29] This led others to try deconstructing film images into a "language" homologous to speech and writing. It involved, in part, identifying similes, metaphors, synecdoches, and metonymies in shots or sequences of shots and building a "syntax" and "semantics" of what are purely visual phenomena.[30] Some film theorists stress the "reading" of films as if they were visual texts (e.g., James Monaco's *How to Read a Film*), but their arguments have been unconvincing.[31]

Unlike writing, which requires actualization by the reader, film as streaming images is actualized at all times. Consider the sentence, *The girl enters the woods,* as read in a novel. The signs call up a sequence of interpretants, along with vague referents. What does the girl look like? How is she dressed? Is she very young or adolescent? How does she enter the woods? Presumably by walking into it. What sort of woods? The original sentence is a syntagm, requiring actualization for the linguistic process to advance.

And its cinematic equivalent? The girl appears and enters the woods. *We see and understand everything.* Metz argues correctly that cinema is not a language (i.e., a system of signs used for communication). Like all art it communicates in only one direction, toward the spectator. The projected image is simply that, never a sign in the sense of standing *for* something else. Completely literal, a cinematic image is a self-signifying sign requiring no interpretant or referent. Not surprisingly, little imagination (the precursor of actualization) is necessary, and misinterpretation is unlikely unless the spectator fails to recognize the syntagmatic progression of narrative. Conventional closed-text cinema is universally popular because anyone can understand it. The spectator's work has all been done. Unlike a novel and its reader, a film can exist apart from its viewer.[32]

Cinema and theater have little in common. Cinema is visual, drama verbal.[33] As proof, consider that a play can be read aloud and understood in a darkened room, but not a film script. In cinema the fictional component is diegetic, in theater mere convention.[34] Furthermore, the space in a play is continuous and exists in real time. By being discontinuous, film space is uninhabitable by the spectator.[35] Nonetheless, we can imagine ourselves there, unlike when viewing a play or zoo exhibit. As Linden writes: "All art transcends the natural standpoint to create an illusory world of emotional depth. It then negates itself as illusion and leaves us with altered eyes to view our everyday world."[36]

The simulacrum of fictive time distinguishes cinema and television,[37] but because the zoo runs on actual time there must be some allowance to suspend disbelief, for the spectator to move sufficiently far outside himself before novel experiences take hold. In such a spectator-controlled environment the passive message is easily subverted and suppressed.[38] At one institution I visited an electronic scoreboard purported to count down the world's remaining species into the year 2050 in accelerated time, implying falsely that today's information can be used to predict future events. However, the deception extends beyond mere inductionism (chapter 11) by suggesting that extinction rates not only are linear but proceed in cinematic time.

Zoo exhibits rank closest to theater—mute, motionless, unfamiliar theater. Plays and zoos rely on props and staged scenes to establish recognizable human spaces and animal habitats.[39] In contrast, cinema takes us directly there, leaving nothing to be imagined. An actor in a play transforms the space by his presence and actions: we recognize his speech and gestures as human. The presence and actions of zoo animals are alien and seldom recognizable, remote both in physical form and the abbreviated behaviors permitted in confinement. To a casual spectator their stature and behavior will forever surpass understanding.

The *event* is the basic unit of all narrative, and this is especially noticeable in films because images impinge on every facet of the story. Where the novel represents (*re*-presents), the film presents; where drama focuses on imitating action, novels and films emphasize development of events. Metz reminds us, "Reality does not tell stories, but memory, because it is an account, is entirely imaginative."[40] As he rightly points out, each event must end before a new narrative begins—that is, a raw new end of the same narrative, freshly cut and ready for splicing into the evolving syntagm. What we see on screen has already happened, but to the spectator both narrative and the events from which it is constructed make it seem immediate.

We attend movies to watch the recounting of a narrative, visual storytelling in a novelistic format. As Metz notes, the telling of the story is so powerful that even visual content is subsumed into plot, and plot is what we remember, except for a few images.[41] Photographs have no such power. More than one is needed to tell a story. In Metz's words: "Yet why must it be that, by some strange correlation, two juxtaposed photographs must tell something? Going from one image to two images, is to go from image to language."[42]

Film is unique in giving an impression of reality and of our perceptual participation. Nothing else does this, not sporting events, art exhibits, plays, zoos, or even photography. Fleeting images are like life; the static scenes in photographs are more illusory because they stand still and let us contemplate them. Metz calls this quality "the appeal of a presence and of a proximity."[43] It is, he tells us, a feeling of credibility. He notes that although films can be divided into "realistic" and "nonrealistic," both forms succeed in making cinematic reality by "imparting to the first an impression of familiarity which flatters the emotions and to the second an ability to uproot, which is so nourishing for the imagination."[44] Reality, he says, assumes presence; only the here and the now seem totally real.[45] Jean Mitry cites Hollywood's use of montage as the technique that created a perfect cinematic fiction. According to J. Dudley Andrew's interpretation of Mitry's remarks, "So great has Hollywood's mastery of this process been that one is often completely lost within the film world because its logic is identical to the perceptual logic of our daily lives."[46] Perfect fictional cinema, in other words, correlates perfectly with total familiarity.

Zoo exhibits stand still, and their inhabitants mostly stand still too, like dysfunctional theater productions in which the actors neither stir nor speak but pose mutely among inedible props. As to cinema, Metz notes, "It is movement . . . that produces the strong impression of reality."[47] Cinematic motion sets objects aside and assigns them the cues we identify with living beings. We understand the dialogue because it speaks to us. In effect, these images become autonomous, assuming lives apart from the actors used to produce them but similar enough to our own that we can accept them as "real" and momentarily abnegate disbelief. An added illusion of reality becomes possible because we interpret films as happening in the present. "Reality" is denoted by visual impressions that in sum make up the diegesis.

Film, like theater and zoological exhibition, is ruled by spatial constraints, but moving images projected onto a screen of finite dimensions have infinite possibilities. To extend space beyond the screen an actor has merely to look off-screen. Sound or music can achieve this too (think of John Williams' distinctive sound track in *Jaws* whenever the shark is near). Noël Burch discusses Akira Kurosawa's film *High and Low*.[48] For an hour or so the scene is the living room of a rich businessman whose chauffeur's son has been kidnapped. Action is confined entirely to interactions among parties involved in organizing a search: the families, police, servants, and so forth. Camera angles shift continuously, perhaps from a scene involving A, B, and C to one in which D joins E. Then C leaves the first group to join D out of view, E having departed while we were looking elsewhere. The characters are interchangeable. By altering visual perspective, space is reinvented in the act of being reused, a luxury denied modernist spectacles.

7

> Don't worry, baby, endings are a conscious thought.
> —Robert Coover, *The Adventures of Lucky Pierre: Directors' Cut*

BORDWELL USED ELEMENTS OF CONSTRUCTIVIST PSYCHOLOGY TO model the perceptual and cognitive processes involved in watching and understanding film.[1] The method assumes perception and thought to be dynamic and goal-oriented. Although simplistic, such an approach produces a useful frame within which to model what a hypothetical spectator experiences.

Mental processing of information can be strictly perceptual (i.e., "bottom-up") or mediated by previous knowledge, anticipated result, and similar "top-down" processes. Both are inferential, relying on premises cued by perceived stimuli and resulting in conclusions based on experience or an internalized system of rules. In the constructivist model the subject actively probes the environment for cues, then filters them through pertinent cognitive machinery as rapid tests of hypotheses. For example, we see the protagonist turn his head, and we hypothesize that either he is looking at his girlfriend or has just noticed a dangerous antagonist approaching. Because his movements are smooth instead of startled our first hypothesis is the girlfriend, and this is accepted or refuted (i.e., tested) by the subsequent shot. It seems to me that what guides this process—Bordwell's *schemata,* or "organized clusters of knowledge"—parallels semiotic referents in which previous experience or perception of an event forms the basis of an interpretant.[2] For example, prior knowledge of frogs mediates a conceptual frog upon hearing the word *frog* or seeing it in print. Bordwell writes:

> After some interval, a perceptual hypothesis is confirmed or disconfirmed; if necessary, the organism shifts hypotheses or schemata. . . . The theory also explains why perception is often a skilled, learned activity; as one constructs a wider repertoire of schemata, tests them against varying situations, and has them challenged by incoming data, one's perceptual and conceptual abilities become more supple and nuanced.[3]

Bordwell then extends constructivist theory to watching a movie. He divides the dynamic into three parts: (1) perceptual capacity, (2) prior knowledge and experience, and (3) material and structure of the film. He gives a succinct summary of how the visual system responds:

> Cinema is a medium that depends upon two physiological deficiencies in our visual system. First, the retina is unable to follow rapidly changing light intensities. At critical "fusion" frequency, more than fifty flashes per second will create the impression of steady light. Second, the phenomenon known as apparent motion occurs when the eye sees a string of displays as a single moving one. This effect depends on the fact that the eye will infer movement from an intermittent input if the jumps are not too large. Flicker fusion and apparent motion illustrate how automatic and mandatory bottom-up processing is: although we know that a film is only a stroboscopic display of fixed frames, we *cannot fail* to construct continuous light and movement.[4]

Once vision has become dark-acclimated, flicker fusion is strengthened. In addition, the darkness blocks out extraneous visual stimuli. Nonetheless, what the spectator sees in terms of a seamless continuum of images is self-created: a film lasting ninety minutes contains approximately forty minutes of black screen.[5]

In Bordwell's model the film-goer taps a reservoir of prior knowledge, or stored schemata, based on everyday experience and a history of going to the movies. Hypotheses are tested and expectations satisfied based on this prior knowledge. I add here that the process is made efficient by standard plots and narratives, which tend to tell the same stories in genres. For example, romantic comedies induce different expectations of plot and narrative than whodunits, thus narrowing the range of testable hypotheses and constraining expectations. According to Linden, "The spectator identifies with the perspective of the camera and his normal bodily image is negated."[6] He participates without joining in, and the exhilaration is in almost being there.

Recent research at the neurological level shows that our brains are hardwired for viewing other human beings and human activities, scenes common to films. We understand them in part because our brains accept them as natural and process them as familiar neural messages. The eye—analogically the displaced subject—wanders freely in a disembodied state across a two-dimensional world. In Vidler's words, "And with the dispersal of the eye, all pretense of

Fig. 19 Clint Eastwood in the 1966 film *The Good, the Bad, and the Ugly*. Whatever he said or did onscreen, the audience processed it with brains in synchrony. MGM Studios.

home, of retrievable places and reconstructed spaces, is gone."[7] Experiments in cognition using functional magnetic resonance imaging (fMRI) demonstrate that when we watch film our own brains are remarkably "in synch" with the brains of the rest of the audience (fig. 19). This is true even to the point where cortical activities correlate predictably at identical segments of the film sequence. For example, the brain's fusiform gyrus is activated when we view images of human faces, the collateral sulcus when the images are outdoor scenes, and the middle postcentral sulcus responds consistently to hand-related movements and hand movements associated with the performance of motor tasks.[8]

In addition, we can identify human faces not merely on the basis of intrinsic information (e.g., eyes, nose, mouth in a conventional pattern), but from contextual clues as well.[9] In other words, even the hint of a human form is adequate to imply the presence of faces when no intrinsic information is available. Certain great artists suspected this before the rest of us (fig. 20). Not surprisingly, these remarkable neurological attributes overlap very little onto other species unless they seem perceptually human: the less they resemble us, the less likely we are to notice them.

Like literature, narrative cinema demands the spectator's partici-

Fig. 20 *Jacqueline Rocque*. (Pablo Picasso, 1957) © 2005 Estate of Pablo Picasso/ Artists Rights Society (ARS), New York.

pation in telling the story: "The film presents cues, patterns, and gaps that shape the viewer's application of schemata and the testing of hypotheses."[10] Comprehension involves identifying causal connections such as plot, organizing and following temporal events, deciding who the characters are and how they interact, and formulating a string of hypotheses and testing them continuously as the action unfolds. In other words, the film spectator, like the literary reader, *constructs* a story in all its intricacies from the cues provided. McConnell explains the difference between literature and film this way:

> In written narrative, we begin with the consciousness of the hero and have to construct out of that consciousness the social and physical world the

hero inhabits. But in film the situation is . . . reversed. Film can show us *only* objects, *only* things, only, indeed, people as things.[11]

This distinction has enormous consequences for zoos. By ignoring narrative, zoos also ignore the principal means by which we move in the world, in particular our reliance on sensory cues to stimulate the intricate neural circuitry that interprets events as stories. We need stories, and we need the narrative eye to be fully open before we can understand them.

Cinematic closed fiction shares with closed literary fiction a *canonical storytelling format*: introduction (setting and characters)—explanation of status quo—complicating action—ensuing events—outcome—ending.[12] This can be modified (or restated) in terms of genre. In romantic comedies the canonical format is: introduction (setting and characters)—explanation of status quo—boy meets girl—boy and girl dislike or mistrust each other, often because of a misunderstanding—boy and girl decide they like each other—boy loses girl, or vice versa—boy and girl meet again and acknowledge their mutual love—happy ending.

We can identify this sequence in the film *Pretty Woman*. Edward is a rich corporate raider; Victoria is a hooker (*introduction*). Edward is tough and insensitive, Victoria sweet and honest despite being a whore, and neither is happy (*status quo*). Edward picks up Victoria as she walks a Los Angeles street and buys her services for a week (*boy meets girl*). Edward comes to resent Victoria's vocation, Victoria is distressed by his insensitivity (*boy and girl dislike or mistrust each other*). Edward takes Victoria shopping and introduces her to high society; she gets him to relax by taking off his shoes in the park and to reconsider the impending raid on his next corporate victim (*boy and girl decide they like each other*). At week's end Edward pays Victoria for her services, and she leaves (*boy loses girl*). In an epiphany Edward takes a limousine to Victoria's apartment just as she is about to return home, having renounced prostitution, and they acknowledge their love (*boy and girl meet again*). Future marriage is implied (*happy ending*).

Bambi, a coming-of-age story, first aired in 1942. As narrative it was conceived and executed in a closed-text format similar to the "formula fiction" of airport bookstore novels, the Travis McGee series by John D. MacDonald, and the *Wizard of Id* comic strip (chapter 5). What this means is that both plot and outcome are predictable, unlike such open-text novels as James Joyce's *Ulysses* and *Finnigans*

Wake, William Faulkner's *As I Lay Dying* and *The Sound and the Fury,* or films like Federico Fellini's *8½* and Christopher Nolan's *Memento.*

The novel contributes narrative structure to film, the theater expression, photography material dimension; together they form the *film esthetic.*[13] The zoo exhibit has nothing comparable. How did we come by our postmodern perspective of film as reality? And how was it that zoos got left behind in the modern?

Starting at the end of the seventeenth century several processes and inventions extended what Friedberg calls the "field of the visible."[14] Significantly, zoo exhibition did not keep pace. Friedberg argues that the shift from modernity to postmodernity has been distinguished "by the increased centrality of the mobilized and virtual gaze as a fundamental feature of everyday life."[15] The meaning of *gaze* in sociological parlance refers to the act of looking as interpreted within a cultural context. Vision obviously keys the gaze, with the other senses (principally sound) being augmentative. This change in human perspective was noted by Benjamin in the gaze of the *flâneur* (literally, stroller) in Parisian arcades during the early twentieth century, but also descriptive, according to Friedberg, of today's prototypical *flânerie,* "the postmodern cinema spectator in the shopping mall."[16] Friedberg's key components of the *flânerie* are visuality, contemplation, and pedestrian mobility. Of their evolution,

> photography brought with it a virtual gaze, one that brought the past to the present, the distant to the near, the miniscule to its enlargement. And machines of virtual transport (the panorama, the diorama, and later, the cinema) extended the virtual gaze of photography to provide virtual mobility.[17]

These diversions, along with winter gardens, museums, exhibition halls, and zoos, constitute what Friedberg terms *protocinematic entertainment,* embryonic mediators of the early tourist's gaze. You could see curiosities of all sorts—ancient artifacts, strange animals, tropical plants, the latest technological developments—all without leaving the city limits. As an interesting corollary, with increasing sophistication of virtuality came decreased mobility until, culminating with cinema, the observer became completely stationary.

It was Baudelaire who romanticized the *flâneur* as a sophisticated male (artist or writer) enjoying the urban life of nineteenth-century Paris with its gas-lit streets, cafés, brothels, theaters—the "perfect spectator," a man with "the love of masks and masquerade, the hate

Fig. 21 Parisian *flâneuse* shopping. Detail from a cover of *Le Figaro Illustré*, January 1891.

of home and the passion for roaming."[18] An urban wanderer, in other words, curious, voyeuristic, hip, aloof.

With the beginning of consumer culture in the mid-nineteenth century came the department store, along with the shopper's gaze (examining goods on display) and the entry of women (the *flâneuse*) into the arcades (fig. 21). At the same time faster modes of travel sharpened and extended the tourist's gaze to distant places. Soon shopping, packaged tourism, and protocinematic entertainment transformed what was now a mobilized gaze into commodities, allowing women to shop, travel, and be entertained without male escorts.

Fig. 22 The Old Arcade, Cleveland, Ohio, opened in 1890. Like other arcades of the era its structure is mainly steel and glass. (Stephen Spotte)

What was initially available only in first-world cities spread around the globe. The modernist arcade (fig. 22) became the postmodern shopping mall. Today the mobilized virtual gaze is not limited to the mall with its cinemas; it's also in every home that has an entertainment center.

In the nineteenth-century arcade, for the first time, the mobile gaze of the *flânerie* was directed through transparent shop windows at endless displays of desirable objects. Even earlier, in the mid-eighteenth century, the shop window had become the "prime proscenium for commodity display."[19] By 1850, sheets of glass large enough to stand from ceiling to floor were being produced, making

Fig. 23 Women safely behind glass. Philipsborn's Department Store, Denver, 1890. (Unknown photographer) Denver Public Library, Western History Collection, X-24030.

possible spectacular displays to entice women shoppers. As typified in figure 23, by 1900, "Female mannequins, posed in static seduction, were women made safe under glass, like animals in the zoo."[20] The *flânerie*—modern sodality of strollers and shoppers, purveyors of the mobile gaze—were ready for the virtual. The shop window as cinema screen (e.g., the window display as *mise-en-scène*, the objects framed and nearby yet inaccessible) follows linearly with a few intermediate stages.[21]

Among the earliest protocinematic entertainments were simple shadow shows (fig. 24) and magic shows based on new technologies. They relied on controlled lighting, stationary audiences in darkened rooms, and optics. The most famous of these made use of the magic lantern, an early version of the slide projector (fig. 25). It preceded the panorama and diorama (described shortly) by at least one hundred years, having been used before audiences in the seventeenth century. Étienne Gaspard Robertson of Liège developed a terrifying show called *Fantasmagorie* that he took to Paris, probably in 1794 (fig. 26).[22] In 1797 he gained access to a chapel no longer in use on the grounds of an old Capuchin monastery, conveniently surrounded by ancient tombs. "Audiences entered through cavernous corridors,

Fig. 24 Shadow show. (Marion 1869, 281)

Fig. 25 *Les spectres*. (Marion, 1869, 313)

marked with strange symbols, and came on a dimly lit chamber decorated with skulls; effects of thunder, sepulchral music, and tolling bells helped set the mood."[23]

In the feeble light of a single lamp and several coal-burning braziers, Robertson denounced charlatans and prepped his viewers for

Fig. 26 Robertson's *Fantasmagorie*. (Marion 1869, 227)

something far better. He threw chemicals (blood, vitriol, and nitric acid) and newspapers onto the fires causing columns of smoke to rise.[24] Then the lamp went out, putting the audience in darkness.

> Onto the smoke arising from the braziers, images were projected from concealed magic lanterns. They included human forms and unearthly spectral shapes. The images came from glass slides, but the movements of the smoke gave them a ghoulish kind of life.[25]

As Erik Barnouw reports, some spectators swooned, convinced they were in the presence of the supernatural.

Images were projected from the rear, which hid the magic lanterns. Robertson also developed a method of projecting them onto gauze that had first been soaked in wax and then ironed to attain a correct state of translucence.[26] These, the equivalent of today's movie screens, were concealed from the audience by a black curtain kept closed until the lamp was extinguished. Robertson also blacked out the areas around the images so they seemed suspended and ethereal. Robertson's lanterns could be moved forward or backward, allowing him to enlarge images and make them appear to approach the spectators until going out of focus, or retract them until they shrank into the distance. According to Fulgence Marion, author of *L'Optique,* Robertson "endeavored, on the other hand, to establish in the eyes of all, the absence of all occult causes and only the action of scientific processes."[27] Perhaps, although profit and fame were certainly motives too. And if education was foremost, he never told anyone how his system worked.

Fantasmagorie ran for six years, but the secrets eventually got out and others copied it. An English version, put on by the magician Paul de Philipstahl, was shown in London in 1801, where it played for two years before moving on to Edinburgh and then to New York in 1803. The New York *Post* noted that the subjects of some of de Philipstahl's "fantoms," which included Washington, Jefferson, Adams, Hancock, and Robert Morris, appeared to have achieved premature "fantomhood" considering that the originals were not yet in the grave. Subsequently, de Philipstahl's promotional literature included the phrase "apparitions of the Dead or Absent."

The results of these performances were not always positive or predictable. In 1806, Andrew Oehler's "ghosts" landed him temporarily in a Mexican dungeon. After being released he gave up magic and settled in the U.S. The next year a Boston performance of *Phantasmagoria* featuring "spectreology and dancing witches" caused a fire in the Columbian Museum, burning it to the ground.

Among other prominent forms of protocinema was the panorama, invented by the Irish portrait painter Robert Barker and patented in 1787. The following year Barker opened the first panorama in Edinburgh, followed in 1792 by the first panoramic painting of 360 degrees in London, titled *The English Fleet Anchored Between Portsmouth and the Isle of Wight*. Others opened exhibitions in Paris in 1800. Helmut and Alison Gernsheim write:

> The general enthusiasm for panoramas in England, France, and other countries was caused by the astonishing illusion of reality of the depicted scene. Placed in semi-darkness, and at the centre of a circular painting illuminated from above and embracing a continuous view of an entire region, the spectator lost all judgment of distance and space, for the different parts of the picture were painted so realistically and in such perfect perspective and scale that, in the absence of any means of comparison with real objects, a perfect illusion was given.[28]

Panoramas were so popular—and so successful as *trompe l'oeil*—that at least one artist took his students there to study Nature.[29] By the early 1800s panoramas had spread throughout Europe and North America. The premier painter of panoramas was Pierre Prévost, and his principal disciple and assistant was Louis Jacques Mandé Daguerre, one of the inventors of photography, whose great talent actually lay in stage design. Daguerre quit working for Prévost in 1816 and accepted a contract from the Théâtre Ambigu-Comique, where he became famous for his sets. Lighting, in particular, was his specialty, and he simulated moonrises, moonlit scenes, even volcanic eruptions and night fog that dispersed at sunrise. His skill in *trompe l'oeil* was unsurpassed. In 1821 and 1822 he also produced designs for the Opéra, directing a moving sun across the stage in the closing of *Aladin* [sic] *and His Wonderful Lamp*.

During these two years Daguerre was also busy preparing the world's first diorama, a specially constructed building in which huge translucent paintings were displayed under the changing conditions of light he had perfected for the theater and opera. He took as his partner Charles Marie Bouton, a distinguished painter of architecture. Daguerre designed a building in the theater district of Paris and had it built. It was a plain structure with long windows placed high to admit light. The exhibit opened July 11, 1822 showing *The Valley of Sarnen* (Switzerland) by Daguerre and *The Interior of Trinity Chapel, Canterbury Cathedral* by Bouton.

The panorama had indeed been protocinema, but the diorama was a major step forward. Spectators entered a dim anteroom and were ushered into a dark circular chamber. The only light came from shrouded lamps placed on the floor to illuminate a few descending steps. People sat on benches before what appeared to be a large window through which could be seen the interior of Canterbury Cathedral undergoing repairs. The workmen were resting. One critic wrote:

> The pillars, the arches, the stone floor and steps, stained with damp, and the planks of wood strewn on the ground, all seemed to stand out in bold relief, so solidly as not to admit a doubt of their substantiality, whilst the floor extended to the distant pillars, temptingly inviting the tread of exploring footsteps. Few could be persuaded that what they saw was a mere painting on a flat surface. (A lady who accompanied the writer to the exhibition was so convinced that the church represented was real, that she asked to be conducted down the steps to walk in the building.)[30]

Spectators gazed at the view under changing light, as if the day had become intermittently overcast. Rays of sunlight appeared, shining through the simulated stained glass windows and illuminating the

Fig. 27 Workings of the diorama. (Marion 1869, 287)

floor with streaks of color. The critic rightly observed that the effect was heightened by the eyes having become conditioned to darkness.

There were noises underneath the floor (fig. 27). Canterbury Cathedral turned slowly out of view to be replaced by the Sarnen Valley with its streams, river, lake, and quaint villages. Again the light changed,

> and the diversified effect produced by the varying shadows, as they become transparent or opaque, according to the approach of storm [*sic*] or the clearing up of the atmosphere, cannot be surpassed. The stillness and clearness of the lake at one moment, yielding at another, as the weather changes, to the successive copper and leaden hues of the dense clouds . . . cannot be too highly admired.[31]

The diorama was a smashing success. Most spectators thought the paintings were models containing real objects, which they tested by throwing coins or little balls of paper at the scenes. Usually two pictures were on view (sometimes three), and a performance took twenty to thirty minutes. The high price of admission limited viewing to the wealthy. On September 29, 1823 a second diorama opened in London to rave reviews.

The large size of the paintings (22 by 14 meters) and the complicated overhead lighting kept the exhibit portion stationary. It was the stage that rotated. The audience was turned approximately 73 degrees until the proscenium lined up with the opening containing the second painting. The seating area could hold forty people occupying nine boxes and 310 standing or seated on the few benches provided.

The paintings were set back from view (thirteen meters from the front row), which concealed their margins and enhanced the illusion of depth. The spectators sat in near-darkness as in a movie theater, waiting for the curtain to be raised. The paintings, illuminated from above and behind, were produced on linen so fine as to be nearly transparent. Gernsheim and Gernsheim write:

> The great diversity of scenic effect was produced by a combination of translucent and opaque painting, and of transmitted and reflected light by contrivances such as screens and shutters. The front of the painting was illuminated by daylight from a ground-glass skylight, into which a number of coloured transparent screens could be interposed to vary the effect. Most of the changing light effects were produced by modifying the daylight passing through the back of the picture (hence diorama—Greek *dia* through, *horama* view) from the long vertical ground-glass windows. This was achieved by interposing a large number of similar coloured screens, which were worked by pulleys and counterweights. In this way the most varied effects from brilliant sunshine to thick fog could be produced.[32]

As Dolph Sternberger points out, "such deception is not meant to deceive but to exist for its own sake, and it is content to amaze the viewer."[33] Although a few gullible spectators believed what they were witnessing was real, the rest were there to experience the very epitome of artificiality. Painted scenes illuminated under shifting light produced changes later used in cinema—what in postmodern times are called "special effects." "Reality" was neither promised nor expected. Friedberg mentions that in 1823 Yorkshireman Thomas Hornor climbed to the top of St. Paul's Cathedral with his drawing materials and telescopes and sketched London in 360 degrees. In 1829 the finished illustration was housed on the roof of Decimus Burton's Colosseum [*sic*]. The fascinating result:

> The rooftop location of this panorama necessitated a new design feature: the first hydraulic passenger lift ("ascending room") carried spectators who did not wish to climb the stairs. The elevator was a mechanical aid to mobility; the gaze at the end of this "lift" was virtual.[34]

Where was the zoo when these other entertainments were evolving? Basically where it is today. The only substantial breakthrough in outdoor zoo design in the nineteenth century was Carl Hagenbeck's moated live animal exhibits, which eliminated the need for bars between animals and spectators. Panoramas at the Dammtor (Hamburg's city gates) had drawn huge crowds. In 1897, Hagenbeck set up *Polar Panorama* on Heiligengeist Field, an exhibit of approximately 929 square meters (10,000 square feet). It combined a painted panoramic scene with a moated zooscape of imitation ice holding polar bears, seals, guillemots, gulls, and cormorants. Nocturnal illumination revealed the background scene—Norwegian arctic explorer Fridtjof Nansen's ship *Fram* frozen in the ice—fronted by the exhibit components. Hagenbeck, who later took his exhibit to the Berlin Industrial Exhibition and then to Paris with the title *La Vie au Pôle Nord*, used these portable displays as prototypes for a permanent arctic exhibit at his zoo in Stellingen, a suburb of Hamburg.[35]

Progress in zoological exhibition since the late nineteenth century has mostly consisted of lessening visual clutter between spectator and animal by continuing to substitute moats for bars in large exhibits, replacing bars in smaller ones with piano wire or glass, or eliminating any sort of barrier in favor of a darkened public area.

Seldom does the visual perspective change. Zoological exhibition has nothing comparable to film's "thirty-degree rule." Film directors in the 1920s discovered that spectators become uncomfortable when sequential shots of the same subject vary less than thirty degrees. Unless a camera angle equals or exceeds this value on the next shot the cut seems hesitant and indistinct from its predecessor.[36] Like other rules this one has been broken repeatedly in film-making, but it shows how surprise, interest, and expectation rely on varied perspective. The attempt by zoos to change visual perspective in a cinematic mode has apparently not been tried. Movie theater designers consider optimum viewing distance to be approximately twice the width of the screen.[37] The entire screen is visible at all times because direct lines of sight from every seat have been included in the view. However, motion at the cinema is limited to moving images on the screen, and the spectator remains stationary. Zoo exhibits in which every part is in view are boring. One solution is to bring visual perspective to the spectator.

Kurosawa's film (see the end of the previous chapter) proves how changes in cinematic perspective are delivered directly from the screen to viewers in their seats. Working in only two dimensions is scarcely limiting to a skillful film director. The same cinematic technique applies to those few situations in which captive animals are in constant motion. Zoos are fond of exhibiting schools of small silvery fishes in cylindrical acrylic aquariums. The effect is pleasing, but never surprising because perspective is invariant, the "actors" always on view. Far more interesting would be an exhibit in which the school momentarily disappears from view, reappears in a different location, swims directly at the spectator, passes in a lateral presentation, and disappears again. Perspective would then change *for* the spectator, just as it does in Kurosawa's film. Moreover, the perspective is cinematic and therefore familiar: the fluid destabilization of space within a fixed frame. This hypothetical situation assumes the same spectatorship of immobility that serves as a necessary condition for cinema. We have still another familiar model of postmodernism, the video game hero who occupies a bounded world of two dimensions. This he traverses regularly, disappearing off one edge of the TV screen and reappearing at another. We seem unconcerned that to connect these paths requires a nonexistent third dimension.

One other use of optics as it relates to protocinematic entertainment and zoo design warrants mention. The English jurist and philosopher Jeremy Bentham (1748–1832) devised the *panopticon* (fig. 28), a model prison constructed mainly of iron and glass. If it seems familiar to zoo professionals the reason will soon be apparent.[38] The building was designed as a cylinder. Prison cells built into the inside curve opened inward toward a central guard tower, to be illuminated by daylight entering through the tower's panoramic windows.

Inmates were invisible to each other but always on view to guards who could see into the cells without themselves being seen. The idea was to separate the captives and deny them mutual influence while simultaneously subjecting them to the threat of constant surveillance. In Friedberg's words, "In the panopticon prison, confinement was successfully maintained by the barrier walls . . . but the subjective changes in the inmate were to be produced by the incorporation of the imagined and permanent gaze of the jailer."[39] Used together, isolation and scopic dominance would alter the prisoners' behavior, rendering them meek and tractable. Philosophically, this represented the apotheosis of the *unseen seer,* "the *seer* with the sense of omnipotent voyeurism and the *seen* with the sense of disciplined surveillance."[40] Bentham later extended the panopticon's potential uses to the classroom, factory, hospital, insane asylum, military

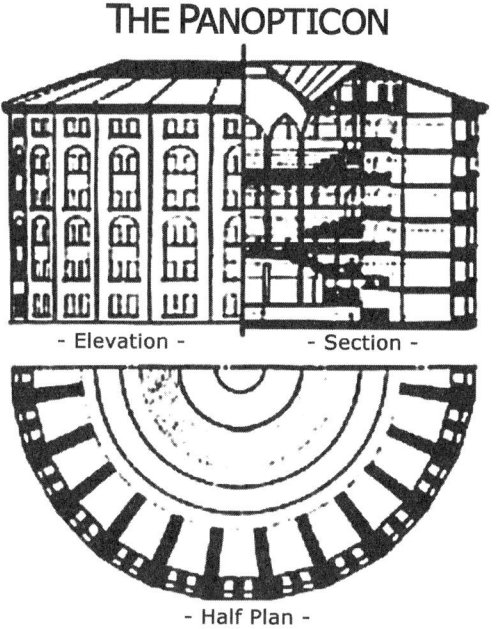

Fig. 28 Jeremy Bentham's panopticon. (Various sources)

barracks—any human institution in which surveillance could be used to dominate those inside.[41]

The panopticon can be seen as a figurative and metaphoric analogy for the cinema simply by thinking of the unseen seer in terms of the "omnipotent voyeurism" of the movie-goer.[42] Actually, the cinema viewer's range is limited by what the film director chooses to reveal. Unlike a view from the tower, what the movie-goer sees is imaginary, and any visual omnipotence is imaginary too.

Bentham's panopticon, although never built in his lifetime, has been copied repeatedly in the design of indoor zoo exhibits. Whether this is done consciously is uncertain. In fact, many zoos make use of a darkened central interior ringed by lighted enclosures into which spectators can peer while remaining largely invisible to the inhabitants.

Packaged tourism, another manifestation of organized spectatorship, appeared in the nineteenth century.[43] Destinations were turned into commodities. As travel became more efficient, tourism "followed a historical development similar to that of the panorama, diorama and cinema where, as the gaze became more 'virtually' *mobile,* the spectator became more physically *immobile.*"[44] Tourists could now see the French countryside or the Canadian Rockies through a train window, what Barthes calls the paradox of "transported immobility."[45] An abbreviated narrative or description of the sites was incorporated into a guidebook or recounted in person by a tour guide, the order of sites to be visited having been prearranged. Zoos have their equivalent in "zoomobiles" and "people-movers," modernist machines designed to transport immobile visitors past the animals. Any narrative or description is once again fitted to the sequence of sites and augmented by a guidebook or tour guide.

The magic lantern induced surprise, fear, and awe, but the panorama and diorama used the image as praxis, eliciting wonder. The viewer, with little effort, experienced the equivalent of Bordwell's *schemata,* or its semiotic counterpart. After all, most scenes depicted either battles (the outcome being common knowledge), landscapes, or the interiors of cathedrals (the last-mentioned known from reading or touring). Such scenes were chosen *because* they were familiar, and each came embedded in a comfortable diegesis despite the absence of narration, perhaps analogous to that of silent film. A common ontology with film seems indisputable.

Observing a zoo exhibit is equivalent to window shopping, today's pastime of the postmodern *flânerie* and their voyeuristic and mobile—but in the zoo's case not virtual—gaze. Thin tendrils still link the nineteenth-century window display with a visit to the zoo or museum in postmodern times, but with a minor difference. Only partly in jest, Susie Fisher asks: "What would you think about a shop that would not let you handle the goods? Are they likely to see the colour of your money?. . . . So what do you think about a museum that puts its objects behind glass and won't let you handle them?"[46]

What zoos provide is a *commodity experience* in the form of a service for which a fee is charged. Karl Marx considered the commodity a kind of fetishism and saw no difference in whether it was satisfied "from the stomach or the imagination."[47] Its exchange value arose in the marketplace, balanced against the exchange value of other objects. He noted that the service provided by a singer satisfies an esthetic need, the joy being inseparable from the voice but a commodity nonetheless.

A zoo's service is to accommodate the postmodern stroller's mobile gaze. The experience, however, remains protocinematic even in postmodern times because service, although a commodity, remains inseparable from the zoo itself; that is, it exists only in the present. The public's interest in sharks was greatly aroused by the film version of *Jaws*, and people flocked to shark exhibits. Clearly, the commodity experience of cinema trickled down to zoos, where the "real" sharks on display lacked both the size and intimidation factor evoked by the fake shark used in the movie.[48]

Eco likens international expositions to the arcade experience on a massive scale—spectacle certainly, but with an element of amusement:

> A boat, a car, a TV set are not for sailing, riding, watching, but are meant to be looked at for their own sake. They are not even meant to be bought, but just to be absorbed by the nerves, by the taut, excited senses, as one absorbs the vortex of projected colors in a discothèque.[49]

We might desire the goods, but then again we might not. It's the show that matters. For the same reason, spectators pay an entrance fee to a zoo where the focus is on captive animals. The way in which the inhabitants are displayed might be irrelevant; that is, whether in tiled rooms with bars or in moated, landscaped vistas. The point is, zoo animals are a commodity available for purchase with an admission fee, mere goods displayed to be visually consumed. The zoo administration, which collects animals instead of stamps or coins or paintings, must somehow, in sustaining a conservationist image, negate all aspects of commodification that accompany ownership. To Benjamin, conferring connoisseur value was sufficient to achieve this.[50] Thus a giant panda, by being both appealing and endangered, is worth more than an ugly lizard that might be just as endangered, and both are worth more than an ordinary cow.

Lack of participation and of seeing oneself in an imagined scene make zoo spectatorship and shopping vastly different. Malls in particular encourage *cognitive acquisition* during which shoppers acquire items mentally and picture themselves trying on or otherwise using products. As Crawford writes, "Armed with this knowledge, shoppers can not only realize what they are but also imagine what they might become."[51] As she points out, the selection of commodities in a commodity-driven society determines self-image, which can be satisfied through consumption. Although the zoo is a commodity, its consumption is limited to the senses. Commodity shopping with its prospect of acquisition, in contrast, is a serious undertaking.

The hyperrealistic experience of shopping is epitomized by the postmodern mall. Among the many marketing strategies is what Crawford terms *indirect commodification,* "a process by which unsalable objects, activities, and images are purposely placed in the commodified world of the mall."[52] We see this in the giant color posters of beautiful people wearing stylish brand-name shoes, the steel drum band outside a music store, or the dolphin show adjacent to a food court. These scenes, which are unexpected to first-time viewers, act as emotional stimulants to encourage spending, a phenomenon called the Gruen effect after the shopping center's principal architect and prophet, Victor Gruen. Hardwick, writes:

> Early in his American career, Victor Gruen realized that the retail environment could entertain Americans better than any show, exhibition, or performance. In all of his designs, he relied on visual surprises to amuse visitors, create consumers, and produce profits. His theory was simple: the more time people spent enjoying themselves in the commercial environment, the more money they would spend.[53]

Gruen designed the world's first shopping mall, called Southdale, which opened in 1956 in Edina, Minnesota. In the history of merchandizing it ushered in the postmodern age. As its architect, Gruen featured a sidewalk café, sculptures, a cigar-store Indian, orchids and magnolias, eucalyptus trees nearly fifty feet tall, a gigantic cage filled with birds, and a children's carnival and zoo.[54] As he correctly predicted, in such surroundings an ordinary commodity is suddenly seen as extraordinary and desirable. Even more important, the mall becomes a destination in itself, a place where pleasant emotional experiences can be repeated At the forefront are those postmodern malls where theme parks and merchandizing blend spectacle, fantasy, and marketing. The marriage seems natural. As Crawford remarks, theme parks are marketplaces in disguise, and malls now entertain. The great American pastime is probably not the zoo, baseball, or even sex. It's shopping.

In the 1930s, Gruen and his associates adapted the European arcade to store fronts on New York's Fifth Avenue. Building a recessed arcade at the entrance and substituting glass doors allowed pedestrians to step "into" the store and away from other pedestrians while examining goods displayed in the windows. In addition, windows lining both sides of the arcade substantially increased the amount of linear display space and gave the illusion of an uninterrupted field of merchandise. We see this technique used repeatedly in zoo exhibits where shallow arcades, or alcoves, allow spectators to step out of the main flow of traffic to examine one or more exhibits.

As scenic entertainment the zoo belongs with other spectacles, which themselves slip under the radar of exegesis except to connoisseurs of curiosities. A zoo exhibit neither transports the imagination from a stationary place of confinement nor offers the notion of a journey. It simply is there, its inhabitants semiotically real objects superimposed on real time. An exhibit containing only a sleeping animal, or one hidden and therefore invisible, is like a film composed of empty sky. The appearance of an actor—anything, even a sparrow just arrived to bathe in the lion's water dish—connotes life and gives the space momentary reason to exist. Now there is *motion*. Nonetheless, still missing are all the crucial and permanent characteristics of cinema—virtual, spatial, and temporal mobility—that allow for transport of the imagination, expansion of place merely by introducing sound or linking images in a montage, and compression of time. Without these the zoo remains trapped inside its outmoded status, viewable solely through the constricted gaze of the spectator.

8

> When we are afraid, we shoot. But when we are nostalgic, we take pictures.
> —Susan Sontag, *On Photography*

A ZOO EXHIBIT COULD NEVER PASS AS ART, WHICH BY DEFINITION functions to make familiar objects unfamiliar, extending the esthetic experience and stirring emotions instead of teaching us facts. The artist directs our gaze not at the object itself but on this other quality until we see the subject for what it *isn't*. As Rudolf Arnheim reminds us, "The perceived image, not the paint, is the work of art."[1] Art can never be explained. A zoo exhibit, in direct contrast, seeks to familiarize the unfamiliar through explanation. The distillation of biological knowledge, which can be transmitted only as information, is art's antithesis.

Zoos make extensive use of color photography for reasons that are unclear.[2] Maybe the purpose is simply decorative, patches of colored light used to brighten otherwise drab corridors. Perhaps the intention is to smooth the rough edges of urban "reality" with photography posing as art. Another possibility is didacticism: photography as explanation, a bridge between immediate exhibit and distant Nature.

None of these hypothetical reasons can be justified. Here and in the next chapter I question the didactic value of photography in zoo exhibits, touch on the emotional effects of photographs in postmodern culture, and evaluate the use of photographs as supplemental embellishments. Much of my thinking has been influenced by Susan Sontag's *On Photography* and Roland Barthes' *Camera Lucida: Reflections on Photography*. Barthes' essay is loose and inductive, less eloquent and analytical than Sontag's, and suffused with an overpowering sadness.[3] It strikes a dissonant chord, reminding us how placing photographic images before strangers often has unexpected effects.

Photographs clutter the postmodern world, distorting and finally overwhelming perspective and immersing us in a thick haze of images. To Sontag their proliferation is clear evidence of kitsch. Sunsets, she reminds us, are now passé because they resemble photographs.[4] Those photographic images displayed adjacent to zoo exhibits compete with others on popcorn bags and billboards and t-shirts, in newspapers and magazines and books, on flyers and posters, in store windows, atop taxis, and tattooed on living skin. Everywhere, a surfeit of photographic images invades postmodernism like crabgrass, crowding us to the edge of daily existence. What *is* a photograph, what limits its functions, and are certain of its attributes antithetical of the soft-focus, harmony-in-life vision of conservation that zoos promote so strenuously? Surely they are if we believe Sontag and Barthes that the camera lens is death's unblinking witness.

Defined naïvely, a photograph is an image produced by a machine. Unlike paintings, which are unique and generated by human labor, photographs can be replicated endlessly, offering interesting possibilities. As Sontag tells us, "A photograph in a book is, obviously, the image of an image."[5] True enough, but the rabbit hole extends deeper. Because of its uncertain ontological nature (the absence of an original), every photograph represents a copy of itself, a quirky metaphysical status not unlike biological clones. In both cases lineage counts more than physiognomy. Two coral polyps with identical genetic composition can look different, just as two photographs made from the same negative (or digital array) might not be identical if one of them has a crease or is printed on different paper. Conversely, although atoms of gold have a common structure, they can never be copies because they are unrelated by descent.[6] When a painting is photographed, an image is produced. It too is an image of an image, although of a separate genealogy.

Photographs are *contingent* because something is always represented, the emergent image being only one expression of many possibilities. To Sontag, "The contingency of photographs confirms that everything is perishable; the arbitrariness of photographic evidence indicates that reality is fundamentally unclassifiable."[7] Furthermore, the object photographed is not merely contingent but immutably so, inhabiting just this specific incarnation. As Barthes puts it, "a photograph cannot be transformed (spoken) philosophically, it is wholly ballasted by the contingency of which it is the weightless, transparent envelope."[8] Any accompanying knowledge is thus momentary and transient, unlike the fixed light of its composition. Events immediately before and after slip into irrelevance. Contrasted with streaming cinematic images, a photograph has neither a past nor a future.

Baudrillard writes, "In fact, it is the eye of the camera that is substituted for time."[9]

What about representation? The idea that resemblance is adequate to represent is false, mainly because pictorial representation is one-way. Goodman and Elgin explain, "A picture represents its subject; the subject does not represent the picture."[10] Moreover, items that show resemblance do not necessarily represent each other (e.g., identical twins). A leopard in a photograph resembles a live leopard, but is not its representation. At the center of representation, Goodman reminds us, is denotation, which is divorced from resemblance.[11] In the case of unicorns and other imaginary animals, a picture is a sign without a referent; denotation is impossible. A picture of a unicorn is not denotative of anything. Finally, because every object in its contingency has protean representations, any statement of resemblance to a picture is inevitably nonspecific.

Barthes considered photographs to be reservoirs of ethnological information. I disagree. He notes that a picture taken by William Klein in Moscow on Mayday 1959 shows clearly how Russians looked and dressed: a boy's cloth cap, an old woman's scarf, another boy's haircut, all superficial moments. Not even this much is apparent in wildlife photography. A snapshot of a rhino reveals little about the surrounding vegetation except the generic shapes of trees or a smear of tall grass, and even less about the animal's skin parasites or its interaction with conspecifics (are there other rhinos outside the frame?). To learn about these things we would need more photographs, a cluster of them from numerous angles and distances, all placed in a particular chronology—a movie, in other words. Even then our understanding of that rhino in the context of the moment could never be complete. Similarly, a color photograph of a lizard placed beside a lizard exhibit reveals no additional information of any use.

Photography's semiotic possibilities are limited, which naturally restricts its effectiveness as a teaching tool. A photograph is inseparable from its referent (in Barthes' metaphor, the referent "adheres"), and the sign is nowhere apparent. Indeed, a photograph is proof that the referent once existed: "The photograph," as Barthes says, "is literally an emanation of the referent."[12] As a result, photographs are not merely contingent, but in their contingency they become tautological. A photograph of a warthog can only depict a warthog that existed in that place and time. Without a discreet referent, a photograph is unable to signify except by becoming something else; that is, by placing a mask over itself and assuming a generalized identity (e.g., slavery, poverty, arrogance).[13] In such situations the mask, or theme, transcends subject matter, but in doing so relegates the object to generic status. We claim to see in these sorts of images symbolic attributes masquerading as social reification.

Above all, every photograph is historical and steeped in melancholy. In snapping a picture the photographer hastens the onset of history and relegates experience to a two-dimensional image on paper. The result, Sontag notes, turns the past into objects perfectly machined for societal consumption. By reducing the scale of experience and then freezing it, cameras "transform history into spectacle."[14] Barthes, who never owned a camera or took a picture, considered the photographic image not in terms of "being there" but of "having been there."[15] He was fascinated by photography's capacity to replicate images of a specific time and place existentially impossible to revisit.

Like Sontag, Barthes detected emanations of the victim in photographed objects: "the person or thing photographed is the target, the referent, a kind of little simulacrum."[16] Each photograph must also be interpreted as the "private appearance" of its own its referent,[17] a condition no less true of an animal photographed during a single moment when it was alive. Placing its photograph beside a zoo exhibit filled with living creatures is mortality's testament, proof that a "timeless" color transparency is actually a death portrait. Not surprisingly, icons of animals are superior to photographs as zoo logos. Icons don't die.

Barthes had a name for these death-in-life images. Every photographed object gives off a *spectrum* "and adds to it that rather terrible thing which is there in every photograph: the return of the dead."[18] In Barthes' metaphysics, photographic images represent eidolons, their ghostly shapes outlined in the streaming photons of reflected light. Photographers were "agents of Death."[19]

What Barthes labeled a photograph's *noeme* ("that-has-been") disintegrates when transposed to cinema: "in the Photograph, something *has posed* in front of the tiny hole and has remained there forever . . . but in cinema, something *has passed* in front of this same tiny hole: the pose is swept away and denied by the continuous series of images."[20]

Barthes senses a "perverse confusion" between the "real" and the "live." To him, "In Photography, the presence of the thing (at a certain past moment) is never metaphoric."[21] By definition, the image and its referent are fused, suggesting that the image had once been alive too. However, also by definition a photograph depicts a moment from the past, a suggestion of death (or impending death). Photography's *noeme* is this: someone has seen the referent *in person*.

Photographs are mute, unable to speak to us except through their captions. Thus an exhibition of beautiful photographs without explanation or description only draws attention to their silence. They can display their beauty like a peacock, but why they exist—and why in this context—will never be known. An uncaptioned photograph, contrary to the dictum, is not worth any words at all. Forever and under every condition photography stands apart from narrative. Despite the claims of some artists and critics, *photography is incapable of storytelling*.[22] The possibilities offered by a painting, however, are less limited simply because art invokes more varied possibilities than the mechanical pattern of absorbed and reflected light offered by photography.

In the end, no form of pictorial representation can incorporate narrative in the manner of written or spoken language. Goodman and Elgin write, "Lexicons and grammars are possible only for systems whose symbols are determinate and discriminable."[23] This excludes pictures of all kinds, which lack rules for identifying and manipulating an established system of signs. In other words, "There is no way to differentiate pictorial symbols sharply from one another, hence no way to determine which symbol a particular mark belongs to or whether two marks belong to the same symbol."[24]

Photographs, Sontag tells us, acknowledge, they do not explain.[25] Furthermore, "Although a photograph may be said to record or show or present, it does not ever, properly speaking, 'describe'; only language describes, which is an event in time."[26] She cites Valéry, who suggested that a writer when describing a landscape or a human face will cause a different image to be conjured in the mind of every reader. The same is true of a photograph, Sontag reminds us. In another context, Valéry's complaint is exactly what distinguishes narrative and invites the reader's participation. Instead of stimulating the imagination, a photograph represses it.

Photography causes us to stand still and look, to bring ourselves to the image instead of absorbing it passively at the movies. Drive down your street as you do each day seeing only a moving panorama, then stop anywhere and notice a particular section of buckled pavement, a gnarled tree, paint peeling on a fence. Are these objects suddenly "real" or merely apparent? As Sontag says, "All that photography's program of realism actually implies is the belief that reality is hidden."[27]

The transformation of reality into images is one of Jameson's two defining conditions of postmodernism,[28] and it should come as no surprise that the overwhelming majority of these are photographs. It was Freudian psychoanalyst Jacques Lacan who, after long study of the emotional displacement and reality confusion epidemic in postmodern times, drew attention to the correlation between schizophrenia and the breakdown of language. The semiotic sign becomes to the schizophrenic more literal, more vivid, until, having lost its interpretant, it is transformed into image.[29] Debord, another postmodern prophet, tells us, "Where the real world changes into simple images, the simple images become real beings and effective motivations of hypnotic behavior."[30] And in case we forget, Sontag offers a jolting reminder: "It is not reality that photographs make immediately accessible, but images."[31] In adding their own images to the chaos, zoos merely raise the level of postmodern noise without ever escaping modernism.

The unique property of the camera is its capacity to alter reality, not in the way art does by showing us something in a new way, but by subverting what we perceive as "real." Objects can be humanized with startling ease. In one famous photograph a vegetable metamorphoses into a nude body builder (fig. 29). The reverse process is no less facile when transforming humans into objects (fig. 1).[32] They are what we see, not what we can perceive, and their presence is all there is.

Postmodern philosophers have detected in photography a form of passive aggression: "This very passivity—and ubiquity—of the photographic record is photography's 'message,' its aggression."[33] Photographs plant themselves in our path, forcing us to look at them. According to Sontag, "To photograph is to appropriate the thing photographed. It means putting oneself into a certain relation to the world that feels like knowledge—and, therefore, like power."[34] Further on, she writes:

> To photograph people is to violate them, by seeing them as they never see themselves, by having knowledge of them they can never have; it turns people into objects that can be symbolically possessed. Just as the camera is a sublimation of the gun, to photograph someone is a sublimated murder—a soft murder, appropriate to a sad, frightened time.[35]

Barthes is no less critical: "The Photograph is violent: not because it shows violent things, but because on each occasion *it fills the sight by force*, and because in it nothing can be refused or transformed."[36]

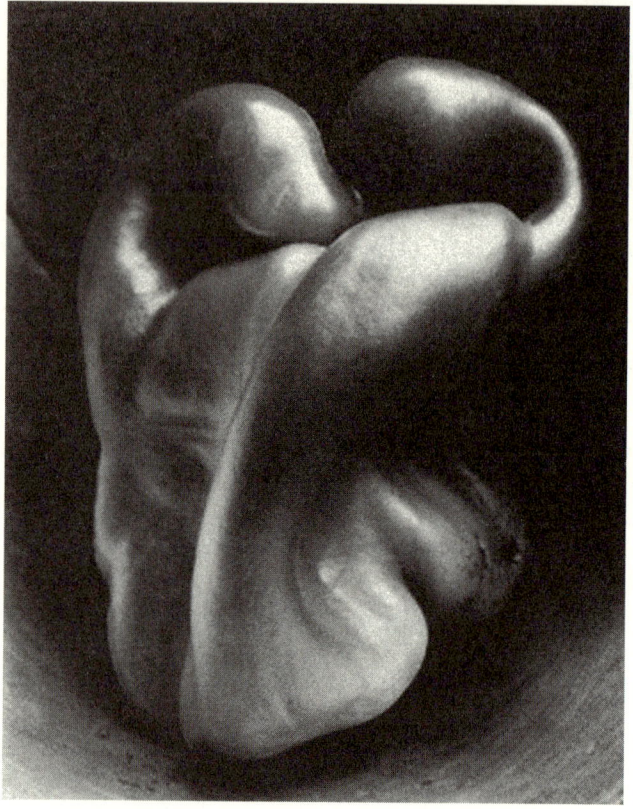

Fig. 29 *Pepper 1930*. (Edward Weston, 1930) Gift of Max McGraw, 1959.665, silver gelatin print, 23.9 × 19.1 cm, © The Art Institute of Chicago.

Barthes uses the Latin word *studium* to describe his general (and passive) interest in most photographs. Rather than "study," he ascribes to the term a cultural connotation in the figures, actions, and settings depicted. Occasionally an image, or fraction of an image, wounds him. This aspect, which disturbs the quietude of the *studium*, he calls *punctum* for its dual meaning of a sting or cut and a cast of the dice. "A photograph's *punctum* is that accident which pricks me (but also bruises me, is poignant to me)."[37] The *studium* is coded, the *punctum* is not.[38] He finds it strange that the "virtuous gesture" brought to bear on "docile" photographs by the spectator—the *studium*—is an idle one: quick glances, desultory lingering, but reading of the *punctum* (Barthes' "pricked" photograph) is quick and alert.

"The incapacity to name is a good symptom of disturbance."[39]

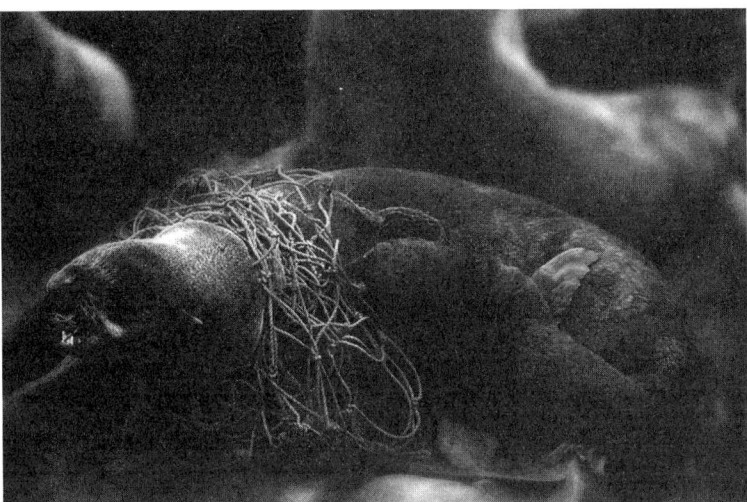

Fig. 30 Net-entangled female northern fur seal and pup. St. Paul Island, Pribilof Islands, Alaska. (Stephen Spotte)

Something bothers us, but what, and why? This is surely not represented in the *studium* if only because we understand it. In a hypothetical photograph of gulls standing on a beach and facing the wind, a fishing lure hangs from the beak of one. We see in the *studium* symmetry of orientation, in the birds themselves a comfortable sameness. The fishing lure—the *punctum*—disturbs simply by being there. It seems cold and dangerous and too frightfully human. The *punctum*, Barthes notes, "is what I add to the photograph and *what is nonetheless already there*."[40] The coded *studium*, in contrast, always speaks for itself, requiring nothing from the spectator. In figure 30 the *punctum* is the section of discarded fishing net entangling a female northern fur seal. It drapes across her shoulders like a death shroud. When it becomes caught on some submerged obstruction and she drowns, her unweaned pup will starve.[41]

Photographs distort, causing us to see ourselves differently. Is it because, during a lifetime of looking at our images reflected in mirrors, we see our bodies turned suddenly around as we appear to others? More likely, photographs betray the tenuous link between emptiness and identity. Mirrors reflect accurately; it is we who reverse the image in our minds and perceive the imperfect symmetry. Or is photographic "reality" simply too strong a contrast with what literary theorists call a vision of the heroic selfhood? Barthes is especially critical and wonders why no photograph of him ever "coincides with my image; for it is the image which is heavy, motionless, stubborn (which is why society sustains it), and 'myself' which is light, divided, dispersed; like a bottle-imp, 'myself' doesn't hold still, giggling in my jar: if only Photography could give me a neutral, anatomic body, a body which signifies nothing!"[42]

Photographs are also records of mortality. In one of many painful, lonely passages, Barthes confesses: "A sort of umbilical cord links the body of the photographed thing to my gaze: light, though impalpable, is here a carnal medium, a skin I share with anyone who has been photographed."[43] Because photographs fail to hide the vulnerability in us, they become, in postmodern times, conduits of sentiment. To Sontag,

> Photography is an elegiac art, a twilight art. Most subjects photographed are, just by virtue of being photographed, touched with pathos.... To take a photograph is to participate in another person's (or thing's) mortality, vulnerability, mutability. Precisely by slicing out this moment and freezing it, all photographs testify to time's relentless melt.[44]

9

> But it is well to be suspicious. Sometimes an image is not an image at all but merely an idea. People have wasted years.
> —Donald Barthelme, "The Falling Dog"

THE ORIGINAL FUNCTION OF PHOTOGRAPHY WAS TO RECORD, AND IN this capacity it evolved eventually into that most demotic of human hobbies. Pretension to art came later. In its early years the daguerreotype was used mainly to make portraits of middle-class patrons, serving as a cheaper alternative to painting. As a method of recording, photography has no equal because the faithfulness of the image is proof of existence. The possibilities of expanding this lone attribute are doubtful.

As art, photography fails utterly. It contains no intrinsic character that identifies the work of individual photographers, nor is any possible when objects can never be pushed beneath the visual surface. So-called "artistic periods" in photography are discontinuous and driven by existing technology. Style is not simply invisible, it isn't there. As Sontag says, "To group photographers in schools or movements seems to be a kind of misunderstanding, based (once again) on the irrepressible but invariably misleading analogy between photography and painting."[1]

I can think of another reason. Why is it, Goodman posits, that "I [can] no more make a forgery of Haydn's symphony . . . than I can make an original of Rembrandt's painting or of his etching *Tobit Blind*?"[2] An "authentic" painting is the original work. Anything else, no matter how closely it resembles the original, is a forgery, or simulacrum. Goodman reminds us that in music and writing there are no such things as forgeries. A manuscript of an original score "is no more genuine an instance of the score than is a printed copy off the press this morning, and last night's performance no less genuine than the premiere."[3]

In Goodman's terms a painting that can be distinguished from its simulacrum is *autographic,* but music and writing are nonautographic, or *allographic,* because no such distinction is apparent. Plays are allographic too, despite differences in stage settings, quality of production, and the actors' deliveries. Neither term applies where an original does not exist, as in cinema and photography, the reason being that arrays of light lack a test of compliance when the image and its origin are moieties. A photograph is instantiated technology derived secondarily from applied physics. Painting, music, literature, and theater are direct instantiations of the mind.

In *Art and Visual Perception,* Arnheim contends that for an object to be transformed into art the human eye must recognize it as a *deviation* from normal visual perception. This is, of course, both a modernist and a postmodern criterion and the end of a long esthetic journey from early nineteenth-century France, where *trompe l'oeil* was thought to be art at its most advanced state. Jack Kerouac recognized the indelible contrast: "Great simple art is always suddenly inexplicable and forever understood; it looms, like the forest."[4] Victor Shklovsky, quoted by Andrew, has this to say:

> The purpose of art is to impart the sensation of things as they are perceived and not as they are known. The technique of art is to make objects "unfamiliar" to make forms difficult, to increase the difficulty and length of perception because the process of perception is an aesthetic end in itself and must be prolonged. Art is a way of experiencing the artfulness of an object; the object itself is not important.[5]

And Sontag writes that "the identification of the subject of a photograph always dominates our perception of it—as it does not, necessarily, in a painting."[6]

Finally, despite how it looks a photograph is no more like art than "found" sculpture (e.g., pieces of driftwood, oddly shaped stones). As Simon Shama points out, these minimalist landscapes remain trapped inside the very criticisms of "unnatural" art they attempt to rebut.[7] Even more insidious is photography's effect on the homogeneity of art. Douglas Crimp explains that once photography enters the art museum, not as a means of keeping records but as art itself, disequilibrium reigns, and the museum as a source of knowledge is invalidated; originality falls to the wayside. "Even photography," Crimp writes, "cannot hypostatize style from a photograph."[8] To postmodern artists like Robert Rauschenberg, photography's malleability seemed limitless: "Rauschenberg steals the *Rokeby Venus* and screens her onto the surface of *Crocus,* which also contains pictures of mosquitoes and a truck, as well as a reduplicated Cupid with a mirror."[9]

If photographs can never be art, what does their use in zoos represent? Benjamin, in his influential essay "The Work of Art in the Age of Mechanical Reproduction," emphasizes that one effect of machine-produced images has been to disconnect service from provider.[10] He uses the film actor as an example. An actor becomes separated from his cinematic image, making him superfluous and commodifying the film itself. Must the actor be present before we can view the film? Of course not, nor the singer for us to hear her songs. As both Marx and Benjamin understood, in an age of mechanical reproduction, goods as commodities have been replaced by services. Traditional reality is no longer needed. Interpreted in this context, photographic images placed beside a zoo exhibit become the primary focus, relegating the animals to secondary status. Think of it like this: we have on one hand an actor (live animal) presented mutely out of context—that is, out of Nature. On the other hand we have a photographic image—an objectified artifact, no less—depicting a lion chasing down a wildebeest, or a hawk snatching a rabbit on the wing (the actor while acting). Such pictures do not so much supplement live exhibits as supplant them. Like an actor, the zoo animal has become subservient to its own image.

Photographs are a template for certain forms of societal authority. In one instance a photographic likeness wedges itself between a subject's presence and its absence. We distinguish a celebrity in person from her "presence" in a photograph, bestowing on the latter an ephemeral reality, although distant with blurred edges. A film buff would naturally prefer seeing his favorite actress up close and in the flesh, and her autographed picture has value to him (i.e., it seems more "real") only because he holds the original in high esteem.

Societal authority prizes wildlife photographs for their "realism." Consequently, the living animal is *not* superior to its image in the minds of consumers, provided its image is photographic. Despite having been cropped, balanced for color, and adjusted for contrast, hue, brightness, and sharpness, a photograph remains, as Sontag says, "a trace, something directly stenciled off the real, like a footprint or a death mask."[11] Even so, in a postmodern world where commodity subverts reality at every turn, "our inclination is to attribute to real things the qualities of an image."[12] A photograph at a zoo is reassurance that the animal displayed beside it is in some manner "real" despite simultaneously disavowing life's vitality.

Postmodernism has bred many social anomalies, among them our addiction to images. The exhibit alone is apparently not enough, although the use of supplemental photographs inevitably reinforces a notion of reality confusion in zoo spectators. Obviously, a semiotically real animal and its artifactual surroundings are inadequate to meet the demand for movement, narrative, excitement. To Sontag, "Photography is the reality; the real object is often experienced as a letdown."[13]

It seems certain that the purpose of zoo exhibits with their absence of narrative is to make the strange familiar, to supplant mystery with the mundane through the imparting of facts. Accordingly, secrets have no place to zoo educators, who seek to drag them forward like ugly trolls blinking back the light. But surely some of Nature's mystery should remain, or if not mystery, at least beauty. What easier way than a lighted box containing a color transparency taken in the wild? A photograph is at once familiar and exotic, never mind that Nature can't be photographed.

A photograph exists for no other reason than to display objects. But even without the skewed vision of an artist to distort shapes and colors, the images lie. The vibrant reds, oranges, and yellows of tropical coral reefs are largely gone at depths beyond two atmospheres; there scenes are drab and suffused with blue unless illuminated by electronic strobes. A dramatic photograph of a bat, wings stopped in flight, is another product of technology. To the human eye, bats are fluttering balls against a darkening sky. And those eye-level shots of insects? Closeup photography has become a cliché, simply another postmodern definition of "hidden" Nature.

We can trace this, Sontag explains, to the Surrealism movement in art. "Surrealism lies at the heart of the photographic enterprise: in the very creation of a duplicate world, of a reality in the second degree, narrower but more dramatic than the one perceived by natural vision."[14] She reminds us that poverty is not more surreal than wealth, or rags than riches.[15] That pop art became Surrealism's descendant was no accident of birth, "For photographs themselves satisfy many of the criteria for Surrealist approbation, being ubiquitous, cheap, unprepossessing objects."[16] As surrealistic images the juxtaposed photographs of an okapi and a cow are now equivalent. And if an exotic example is required, what could be more surreal than a photograph taken in life of a thylacine (fig. 31), a creature hunted to extinction a mere seventy years ago?

Fig. 31 Captive thylacine, among the last living member of its species. (Unknown photographer)

To the Surrealists, the "real world" not only lacked reality, it had become platitudinous. What mattered instead was identifying the locus of visual perception, and this could best be achieved by diminishing Nature. Magritte accomplished it in a series of paintings named *Man's Fate* (fig. 14). They show an easel with a painting of a natural scene against a window. Each painting is apparently finished; the artist has walked away. However, the work on the easel is perfectly contiguous with the view outside: the edges of clouds line up exactly, a ripple on the sea's surface carries without distortion across the canvas.

Reality, Magritte seems to be telling us, is inadequate, and most wildlife photographers would agree. Examine the photographs in a zoo brochure and notice the absence of bars and moats. Consider the many photographs taken of zoo animals at high shutter speeds (often with strobes) or small aperture openings for the purpose of eliminating walls and other evidence of confinement. Think of attempts to make captive animals appear "wild" by framing and cropping pictures to render the trappings of captivity invisible. It isn't the camera that lies, it's the photographer, and the result is often believable. In his autobiography *Chronicles of Wasted Time,* Malcolm Muggeridge emphasizes photography's insidious skill for propaganda and the fostering of instant myths: "The camera has spoken, and before it tongues and pens are powerless."[17]

A display of photographs in a zoo tells spectators, this is what the world is *really* like. Because the images themselves are unable to communicate, zoos give the impression that Nature is like a photograph. *National Geographic* and *Audubon* magazines reinforce this notion until images have replaced what can be seen, smelled, or touched. The postmodern standard of beauty has been reduced to photographic images, whether the subject is a woman, a can of frozen orange juice, or a bird. The captive animal is only semiotically real while the photographs nearby have come to represent reality. The camera no longer simply records, it beautifies, and in doing so changes how we view and interpret our surroundings.

Adjacent to an exhibit of rainforest frogs stands a series of color transparencies of similar frogs in the wild. Uncaptioned, they tell no story. The living frogs are tiny and cryptic, those in the photographs large and in full view. What, exactly, is the message? There is none. Knowledge and understanding are accessible only through narrative. Any information delivered by the graphics is muddled by pictures without words. Sontag writes:

> Photographs, which cannot themselves explain anything, are inexhaustible invitations to deduction, speculation, and fantasy.... Photography implies that we know about the world if we accept it as the camera records it. But this is the opposite of understanding, which starts from *not* accepting the world as it looks.[18]

Neither is the matter of reference clarified. A photograph of a rainforest frog denotes its subject if we consider this specimen to be an individual; otherwise, it denotes the general class of frogs, the same as a picture in a field guide of South American amphibians.[19] Unlike abstract art, photographs can exemplify. A semiotic sign that instantiates its referent simultaneously exemplifies it. Thus a photograph of a frog, useful in a book where it stands alone, is redundant in a zoo exhibit.

Zoos have not yet realized that wildlife photography has become banal, that a sense of reality diminished by repetition anesthetizes the conscience and subsequent concern for Nature. In the case of zoo photographs only the beautiful need apply, the criteria for selection having themselves become platitudes: "Contrary to what is suggested by the humanist claims made for photography, the camera's ability to transform reality into something beautiful derives from its relative weakness as a means of conveying truth."[20] The path is easy: simply follows standards established by others. Thus it is that *trompe l'oeil* has its postmodern incarnation in the wildlife photograph placed to compete with a living animal.

Judging photography on the basis of perfect lighting and composition, sharp focus, excellent balance of colors, and so forth attests to an anachronistic method that Sontag calls *photographic seeing,* or expanding the conception of what passes to define photography as art.[21] At its worst what is vulgar in the extreme becomes transcendent in the extreme with the assumption of a mask, or theme, and therefore profound.

When is a photograph "good?" According to Barthes, when it induces us, vaguely, to think. "Ultimately, Photography is subversive not when it frightens, repels, or even stigmatizes, but when it is *pensive,* when it thinks."[22] Pensive images are rare in zoo photographs, where static beauty is paramount. A beautiful wildlife photograph can never be pensive unless the defining qualities of its attractiveness are combined with an equally strong alternative quality, one that disturbs our equilibrium. The photographs displayed in zoo graphics and in glossy magazines are what Barthes would label *unary:*

> The Photograph is unary when it emphatically transforms "reality" without doubling it, without making it vacillate . . . no duality, no indirection, no disturbance. The unary Photograph has every reason to be banal, "unity" of composition being the first rule of vulgar . . . rhetoric.[23]

Photography fractures our vision, restricting objects to what can be imprisoned inside the frame, excluding from our protophotographic minds access to larger or smaller wholes. As a scientist doing field work I learned to put the camera aside if I really wanted to *see*. This is hardly a new observation. Sontag explains that "the habit of photographic seeing—of looking at reality as an array of potential photographs—creates estrangement from, rather than union with, nature."[24] Further, "Photographic seeing, when one examines its claims, turns out to be mainly the practice of a kind of dissociative seeing, a subjective habit which is reinforced by the objective discrepancies between the way that the camera and the human eye focus and judge perspective."[25]

Can photographs reveal their power metonymically? I suppose so,

Fig. 32 *Wandering Violinist, Abony, Hungary.* (André Kertész, 1921) Gift of the André and Elizabeth Kertész Foundation, Image © 2005 Board of Trustees, National Gallery of Art, Washington, D. C.

although in wildlife photography this would be difficult. To human beings one deer looks too much like another. We detect in its features only what we put there: sadness, fear, worry. Nothing like the texture of a dirt road in Hungary identified by Barthes in André Kertész's photograph of a blind gypsy violinist being led by a boy (fig. 32). But where lies the metonymy? What Barthes perceives, by his own admission, is the referent, the nostalgia of having been there himself—a memory. The road, Barthes admits, is not evidence of the photographer's art, merely his presence: how could the road have been separated from the violinist? "The Photographer's 'second sight,'" as Barthes says, "does not consist in 'seeing' but in being there."[26]

A photograph, then, is a tracing, but its eerie unreality results from its very stillness. Metz reminds us: "The strict distinction between object and copy, however, dissolves on the threshold of motion. Because movement is never material but is always visual, to reproduce its appearance is to duplicate its reality."[27] Perhaps. Zoo animals share with those in photographs an allegorical fragmentation, the one a shard of an ecosystem, the other a remembrance of its own demise.

10

> To perceive the aura of an object we look at means to invest it with the ability to look at us in return.
> —Walter Benjamin, *Illuminations*

WHERE SOME SEE EMPTY SPACE, OTHERS SEE SPACE FILLED WITH emptiness. Where an urban dweller thinks of a thicket as vegetation in need of pruning, a naturalist envisions wildlife habitat. What aspects of visual perception influence our capacity to distinguish objects in spatial dimension, and how does this affect the esthetic postmodern eye? Here I identify several supporting structures of visual esthetics and apply them to the spatial elements of zoo exhibits. I contend that despite efforts made to simulate Nature, esthetic appeal in most zoos is limited even in a modernist sense by the misrepresentation of space. Not taking into account those factors known to direct the spectator's gaze and influence visual perception quells any positive impact of the intended message, leaving feelings of apathy and confusion.

Edwin A. Abbott's science fiction novel *Flatland,* first published in 1884, describes a planet of two dimensions populated by inhabitants whose angular geometric shapes occupy space like sketches on a sheet of paper. Without a sun to cast shadows even the light is flat, and everyone appears to everyone else as a straight horizontal line, either moving closer or receding. Life is dull, Abbott's unnamed protagonist complains: "How can it be otherwise, when all one's prospect, all one's landscapes, historical pieces, portraits, flowers, still life, are nothing but a single line, with no varieties except degrees of brightness and obscurity?"[1]

Life on Flatland has no *perspective,* that capacity to see, draw, and imagine three-dimensional images.[2] The existing light, we must assume, originates somewhere, but the light that falls on Earth gives human beings the capacity to see space in 3D (this and our eyes and central nervous system, which have evolved to use it). As Gyorgy Kepes explains:

Motionless objects . . . are perceived as though flat, and only when the eye receives a successive flow of light patterns reflected from an object can we recognize depth and detect the object's three-dimensional extension. The changing position of our eyes relative to an object reveals its characteristic three-dimensional "thingness."[3]

The protagonist of *Flatland* is a common square. One day a sphere arrives. He yanks the square from his plane life[4] and introduces him to the concepts of "above" and "below," along with the possibilities of existence in three dimensions. The square is astonished, his perception changed forever. The lesson is that zoo keeping might not be controversial if humans were flat-sided geometric shapes living on Flatland. But as the equivalent of postmodern spheres we carry perspective with us; therefore, we intuitively understand spatial dimensions.

Viewing animals inside an enclosure bounded by barriers agitates many people simply because the perspective seems wrong. This remnant of modernism is perhaps the most cathectic concern and source of the sometimes intense criticism of zoos. Consequently, the concept of space leads directly to the dilemma of confinement and the troubling notion that undeserved punishment has somehow resulted in jailing of the mute and innocent. If zoos are indeed "arks" insulated against the raging flood of humanity, they are also floating brigs. In a world where prospects of survival grow dimmer each day for both human and beast, a few endangered species are likely to persist only in captivity.[5]

This vague feeling—and it is a feeling, a Peircean Firstness when abruptly confronted—is implicitly confirmed each time a zoo's advertising uses photographs staged to make bars and barriers disappear, or taken with wide-angle lenses to expand the field of view and exaggerate depth perspective. Whether noble or disingenuous, the "ark" metaphor capsizes quickly if captive animals are perceived as inmates, their keepers as prison guards. Those who advance zoo culture then do so at the expense of victims putatively incarcerated for their own good, collateral damage in humankind's assault on Nature. Sontag writes that during wars, photographs of dead victims "create the illusion of consensus."[6] With zoos the objects were once alive, reiteration of the "ark's" beneficence. All supposedly is well because the proof is in the pictures. But we know from earlier discussions how photographs are actually death's assiduous herald.

The zoo animal as ecosystem fragment demands a spatial scale, especially if displayed in some sort of "immersion" exhibit. The first problem is one of recognition: realistically, within the scale of an ecosystem, is such a fragment even recognizable? The difficulty, of course, is compounded by orders of magnitude when exhibiting elephants compared with mice.

The second problem is context, specifically the context of fragments. Vidler addresses this in terms of Toba Khedoori's paintings. He writes that "a fragment demands a context, a possible and easily visualized site from which one might imagine it was initially snatched, and to which it might, just as easily, be envisioned as returning."[7] Implicitly, "immersion" exhibits are as much about repatriation as snatching from the wild.

Third, as Vidler notes, "the common denominator of all historicist fragments [is] their place in and implications for a narrative of some kind or another, a story, so to speak, of which they formed a part and which they encapsulated in some symbolic or allegorical sense."[8] But zoo exhibits, like Khedoori's paintings, typically resist narrative in favor of proselytizing. Around and outside the exhibits themselves, overwhelming visual clutter in the form of semiotic signs and simulated objects tends to destroy any sense of place.

Architectural space in zoo exhibits is never liberating and often dystopian. This makes the first problem seem the most contentious, especially in light of modernist theory, which dictates that space and time are both destructive of objects.[9] We find zoos and the objects they contain preserved in Vidler's "entropic stasis," a defined space like cooled lava where all movement has stopped. Natural space, in contrast, occupies a gradation from close confinement to open vista. Along this continuum are Nature's living representations, each inhabiting its specific range. Even the Arctic tern, which annually migrates between the edge of the Antarctic icepack and tundra north of the Arctic Circle, a distance nearly equal to Earth's circumference, occupies flyways in three dimensions.

We tend to think of animals stereotypically in the context of how we ourselves perceive their spaces: moles in burrows, bears in dens, monkeys in trees, zebras on savannahs. These notions then carry over to zoo exhibits, where a monkey climbing the inside of its wire enclosure or a zebra standing in a paddock seems to occupy a skewed space that jars us visually. We must then ask, is it the space itself that seems inadequate or some other element of confinement? A large cage might be big enough for a small monkey; perhaps what pains us is watching it climb the barriers of its enclosure.

Enclosed spaces are distinguished by structures such as walls, colonnades, trees, shrubs, levees, glass, or bars. I shall refer to them collectively as walls. Our perception of whether a space seems intuitively enclosed or open is influenced less by its actual size than by other variables. Two of these are wall height as viewed from the front and actual distance between wall and observer.[10] In other words, the ratio of wall height to wall distance (H/D ratio) is an important determinant of visual perception along the enclosed-openness continuum. This is true regardless of scale or actual size of the space.[11] Specifically, the perception (or impression) of enclosure intensifies as the H/D ratio increases (i.e., as the horizontal distance to the wall diminishes in proportion to wall height).

As shown in figure 33, when the height of a façade (i.e., the back wall of an exhibit) equals the horizontal distance from the spectator to its base, the cornice is at 45 degrees from the forward horizontal line of sight. The upper limit of our forward field of view is about 30 degrees, making us feel enclosed. When the façade's height is half the horizontal distance, it coincides with the 30 degree upper limit of vision. According to Paul D. Spreiregen, "This is the threshold of distraction, the lower limit for creating a feeling of enclosure."[12] At a third the distance, we see the façade at an angle of about 18 degrees. Spreiregen explains:

> At this proportion we perceive the prominent objects beyond the space as much as we do the space itself. When the facade height is one-fourth our distance away from the building . . . we see the top at a 14° angle, and the space loses its containing quality and peripheral facades function more as edges. The sense of space is all but lost, and we are left instead with a sense of place.[13]

This last was probably what Hagenbeck unknowingly sought with his moated exhibits. More recently, virtual reality games that exist in the hyperreal make use of these same principles to establish illusions ranging from claustrophobic to agoraphobic.

In practical terms, zoo exhibits can be planned to appear less confining regardless of the actual space available. One way of doing this is to effect a mental switch, focusing less on how the animal might be perceived in the context of its space than how spectators are likely to perceive the space itself. This information, which zoo architects might claim to be obvious, is rarely reflected in their results. In fact, the first empirical test of how the H/D ratio affects human visual

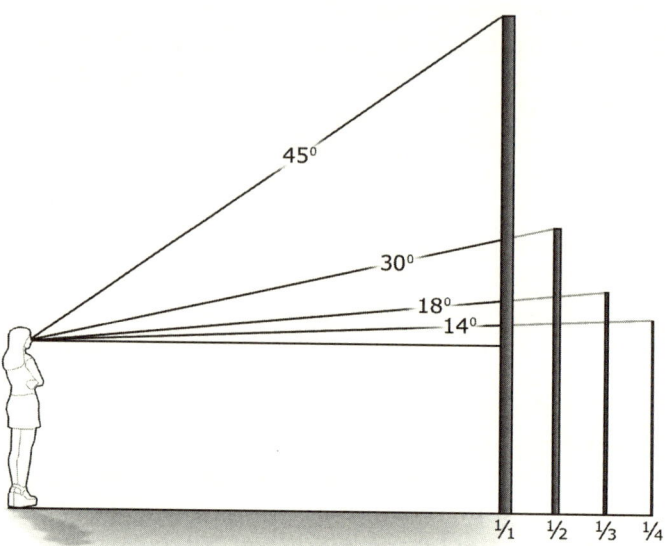

Fig. 33 Relationship between height of a façade and its horizontal distance on the spectator's feeling of enclosure. Angles are approximate, proportions not to scale. See text for details. (Stephen Spotte)

perception was published in 1974 by Scott C. Hayward and Samuel S. Franklin. Their findings: "A 100 square foot (ten x ten) space and a 1600 square foot (forty x forty) space yielded almost identical impressions of enclosure, providing their boundary wall size-distance proportions remained equal."[14] What matters in terms of human visual perspective is obviously not the physical space, but how it is arranged. What this same space means to the animal, of course, is another matter.[15]

The exterior shapes of buildings influence whether we think of them as large or small. Cheyne L. Bamford studied volume perception of 3D structures viewed from an exterior perspective.[16] He found observers to consistently overestimate the volumes of rectangular forms, compared with square ones, and this effect increases with increasing rectangularity. Structures with flat roofs are judged to be larger volumetrically than those with shed or gabled roofs. Volume estimates increase with decreasing distance between observer and wall, and tall structures appear to contain larger volumes than shorter ones having equal actual volumes. These results also bear directly on the design of zoo exhibits.

In 1979, Benedikt conceived and described a novel concept of architectural space. He named its basic unit an *isovist*, defined as "the set of all points visible from a given vantage point in space and with respect to an environment."[17] An isovist thus "defines a field of vision from which various geometrical properties, such as area and perimeter, can be calculated."[18] An *isovistic field* comprises isovists from all visual vantage points in a predetermined environment, whether a room, gallery, zoo exhibit, or composite of city blocks. What architects call "sight lines" are comparatively inexact and incomplete by not comprising *fields* of projected lines and shapes in 3D.[19]

To my knowledge the isovist concept, which is also being used to define space in virtual reality, has yet to be applied in zoo design. By being "location-specific patterns of visibility,"[20] isovists can serve as a principal design tool for computing all visible locations in individual exhibits or throughout an entire zoo, eliminating accidental obstructions, enhancing visibility at important locations, and replacing tired exhibit forms (typically squares, rectangles, cylinders, ellipses) with interesting polygonal shapes and unusual spatial perspectives. Zoos will never escape the restrictions of modernist design, but this could be an important step forward.

We should be worried, standing under a relentless rain of images and stalked softly by that timely predator, virtually reality, never considering that postmodern architecture might, according to Vidler, be situated "somewhere in the space between multiple screens."[21] Fortunately, we tell ourselves, our senses know the difference between reality and representation, and so we sort through the visual clutter hypostatizing what we see without realizing that all mental images are illusions, shadow shows on the walls of Plato's cave. Philosophers and cognitive psychologists call this *naïve realism*—a belief that the everyday world mirrors the actual physical world. In fact, our contact with phenomena outside the senses is purely representational. Reality exists on another plane, if it exists at all.[22] And anyway, who decides the boundaries of reality? As Hayles admonishes, "If every species constructs for itself a different world, which is the world?"[23]

The most challenging problem might be how visual perception allows us to understand visual space, apparently an acquired skill.[24] To some extent our deficiencies in this area can be compensated by the brain, which "fills in" blank spaces so that what we perceive is, in the end, more complete than what we see. This ability relies on previous experience ("top-down" knowledge). Take, for example, the Dalmatian in figure 34, whose outline the brain makes whole by

Fig. 34 Visual perception pieces together an incomplete image of a dog. (R. C. James)

sketching in the missing parts until what we perceive is a finished image.[25] In *externalizing* the world, we experience the dog's image as being external to ourselves. We see space in three dimensions but experience visual events sequentially. The world of our perception therefore exists in four dimensions, 3D space with the addition of time.

Besides light and color we perceive surfaces, textures, and gradients. Elements of texture are large and well separated when nearby, but with increasing distance they appear smaller and more densely packed (fig. 35).[26] From the vantage point of zoo exhibition this means that all important detail and surface relief needs to be close to the spectator, and what appears in the background is less important. Sometimes the proper distance effect can be achieved simply by using dimmer lighting along the back perimeter (fig. 36).

Painting teaches us that balance, shape, and form can be blended and the result hung vertically for viewing in two dimensions. From the start, painting subverted the ordinary and trained the artistic eye for shadow shows, panoramas, dioramas, and films that came later. Balancing all visual elements of a scene not only enhances esthetic

Fig. 35 Background objects have greater visual density. (Stephen Spotte)

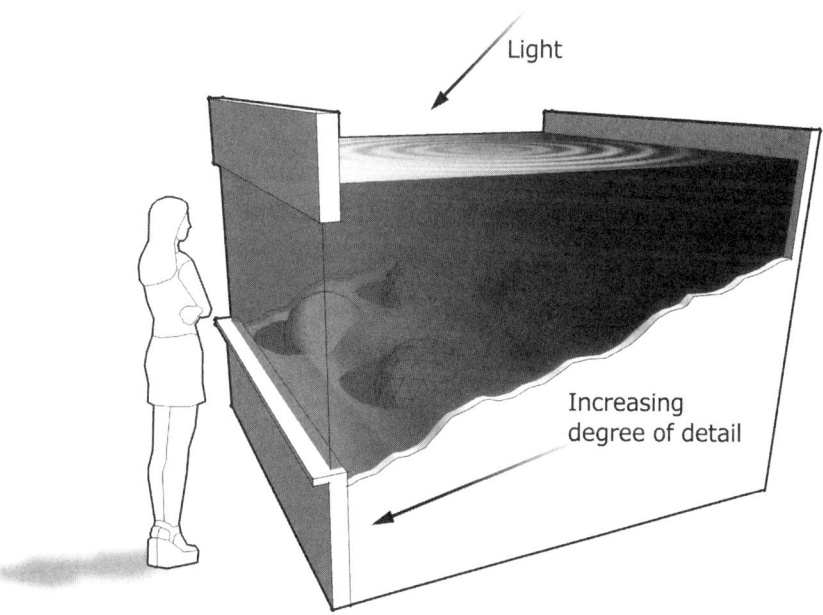

Fig. 36 Detail can be seen only in adequate light. Exhibits with gradations of light give the impression of extended depth. (Stephen Spotte)

appeal, it removes distractions that might block out the message. Certainly we postmodernists should have learned this, but pockets of anachronism remain, of which zoos feature prominently. An exhibit is framed like a painting, and its composition is subject to the same criteria of balance and shape.

Gravity controls our visual impressions, anchoring things in their places. We grow up thinking of mountains spread across broad bases, not tilted over and teetering on their peaks. Trees extend roots downward, not toward the sky. We navigate daily through a maze of stationary objects without the least discomfort, expecting things to stay immutable. As individuals, we move freely about without staggering or falling down because evolution has tuned the balancing mechanism of our inner ears, letting us live in equilibrium between land and sky.

For an object to be visible it must occupy space, but once in place its positioning with respect to the viewer's innate sense of gravity and equilibrium—that is, its *balance*—determines how we perceive its esthetic qualities. Importantly, balance influences any effect on us, including whether we walk away from a zoo exhibit or painting with a favorable impression. Think of balance in terms of Newton's third law of motion: for every action there is an opposite and equal reaction. What can be pushed to one side is balanced by pulling from the other side. Balance is not symmetry. A painting can be balanced yet retain major asymmetric elements.

In Arnheim's metaphor, a balanced picture is one that has reached its lowest state of energy, and all action has come to a stop, calling up entropy and the second law of thermodynamics.[27] Not only does further movement seem impossible, but all parts of the whole appear "necessary" to hold the composition together. An unbalanced composition, in contrast, "looks accidental, transitory, and therefore invalid."[28] The placement of objects or elements of color seems haphazard, distracting the spectator and drawing her gaze *away* from what should be the principal focus. Significantly, "the artistic statement becomes incomprehensible."[29]

Arnheim holds that the *weight* and *direction* of visual objects influence balance. Unlike gravity, which pulls downward, visual weight can be directed anywhere, so long as it connects the spectator's gaze with the intended object. Think of this connection as an invisible *line of tension*.

Several factors affect weight, one being *location*. Typically, the strongest position, or position of most weight, is near the middle of a composition. This *center of focus* can then be counterbalanced using smaller objects distributed toward the periphery. In addition, the weight of a visual object or element increases in proportion to its distance from the center.

Weight is also affected by *spatial depth*.[30] Generally, the greater the *depth of field* (as in vistas) the greater the weight. *Size* counts too: the larger the object the greater its weight. *Colors*, like light, are elements instead of objects. Red is heavier than blue, and bright colors weigh more than dark ones. Because of differential reflectivity, a black area must be larger than a white one for them to have equal weight (i.e., the brighter surface looks larger). *Intrinsic interest* is another factor, directing the spectator's gaze in the direction of particular objects or elements. In zoo exhibits the animals ordinarily serve this function. *Isolation* adds weight by reducing visual distraction, *shape* by assuming geometric forms that seem heavier than free forms. *Balance* is therefore the attainment of stasis, of a visual freezing of objects and elements and the recognition that further shifting is unnecessary.

The *direction* of visual forces is controlled by several factors, including the *attraction* applied by the weight of objects or elements nearby. An attraction can work forward, backward, or to one side. Accompanying a visual distribution of weight is an illusion of motion. This can be seen in Edgar Degas' *Peasant Girls Bathing in the Sea at Dusk* (fig. 37). Attraction is toward the background as the girls lean left and forward, pulling each other into the sea. The girl on the right, although facing away from the setting sun, is nonetheless leaning in its direction. The overall impression is of directed motion.

Visual weight in this painting is provided by the sun, which is the center of focus: orange (a heavy color), the brightest object in view, and positioned within a vista. The girls, who are bathed in shadow, are not in competition for light despite being foregrounded. Instead, they counterbalance the sun with the slender, oblique lines of their figures, drawing it downward and nearer the center and suffusing the scene with added weight, that of light reflected off the sea.

Arnheim believed patterns of balance to be important, referring to a *hierarchic gradient* that approaches zero when a pattern contains many objects or elements of equal weight. Skilled artists avoid such situations, but we observe them commonly in zoos. In a typical display of arrow-poison frogs the mass of objects—vegetation, pieces of wood, rocks, moving water—often renders the intended center of focus (the frogs) at least trivial, if not invisible.

We learn early that objects have specific shapes, a visual form and

Fig. 37 *Peasant Girls Bathing in the Sea at Dusk.* (Edgar Degas, 1875–1876)

texture allowing us to distinguish them from other shapes, and we archive them in a growing library of memory. We recognize a deer by its characteristic shape, trace its outline, recall its texture. Each of us sees differently, but we all notice a salient feature or property of the object observed. We recognize books by their rectangular shape, automobiles by their relative size, metallic appearance, and movement. In the distance, a trotting animal with its nose to the ground is more likely to be a dog than a cat. Young children learn to discern "doggishness" before they can distinguish one dog from another.[31] The heavy, plodding walk of cattle is a bovine trait. We come to know these things and recognize them intuitively, and incomplete images are completed by visual perception and imagistic memory acting in concert (fig. 34). Many zoo animals, however, are outside everyday experience; that is, no images of them have been previously archived in the mind. To know them we must learn their shapes and textures. This is made more difficult if the surroundings detract and confuse.

11

> If you start from facts, the simplest fact requires reasons which are far too complicated, and if you first lay down principles, you begin with the absolute, faith.
> —Gustave Flaubert, *Bouvard and Pécuchet*

I END THIS ESSAY WITH A FEW WORDS ABOUT SCIENCE. IT SEEMS CLEAR that although zoos are stuck permanently in the modernist era they express postmodern sensibilities where animals are concerned. Zoos claim to be scientific institutions while sometimes projecting captive animals as pets or errant children, revealing an incomplete understanding of what constitutes science and its methods used to describe the world. Often the viewpoint is one of paternalistic oversight in the interest of Nature, the zoo laboratory as the "ark's" official sextant.

Evidence for my contention can be found in membership publications, in presentations by staff members to zoo spectators (e.g., bird shows, dolphin shows), in exhibit graphics, and in TV shows about animals filmed at zoos. In these situations an animal might be described as happy, uncooperative, stubborn, not feeling well, depressed, or missing its stable mate. Of course, no one has ever identified these emotions using scientific methods, and they seem especially improbable to anyone who admires Nagel's supposition about the impossibility of inserting ourselves squarely into the life of another species, or even another human being. As constructed generalities, terms like happiness and depression are without objective meaning, only vaguely descriptive, and peculiarly postmodern in the cavalier disregard of what we actually don't know about animals.

Most zoos maintain records on behavior, health, feeding habits, and sociability. They describe these activities as scientific work, and no doubt many spectators and TV viewers believe them without realizing that science is *always* linked with theory and that all data lack scientific value except in this context. Notes and records might be useful for husbandry purposes, but husbandry is not the practice of science.

Keepers and veterinarians are presented as authorities on captive animals and their preservation. Perhaps greater modesty is warranted considering that there *is* no higher authority on the subject of knowledge, and if truth resides anywhere its habitat has yet to be revealed. The result of authority-seeking can only be *infinite regress,* or the tracing of a *lineage* of authority, which leads not in the direction of progress, but backward.

We know for certain that knowledge can't be gained by observation or logic. Put more forcefully, knowledge has no source at all, at least none discernible to us. Popper showed how the empirical approach to knowledge leads nowhere.[1] If an assertion is made a critic will ask for justification. She will ask, in effect, for the *source* of the assertion. No matter what the answer, a still wider search will be necessary. Offering a literature citation, for example, can only lead to the question, *But how does that author know?* We can then check the sources cited by this authority, but the trail always meanders in reverse, disappearing into a thicket of infinite regress. Even a documented observation—the ultimate answer to a question posed by an empiricist—is empty of finite knowledge.

The empiricist's questioning of our source of knowledge is misplaced by requiring an authoritarian answer, which ultimately is no better than any other answer. A question thus posed, as Popper observed, is genetic: "It asks for the origin of our knowledge, in the belief that knowledge may legitimize itself by its pedigree."[2] Perceived authority has no bearing on truth. Pure knowledge, if it exists, originates from metaphysical sources, placing it outside the scientific method. As Eagleton says, "The world is not an object 'out there' to be rationally analyzed, set over against a contemplative subject: it is never something we can get outside of and stand over against."[3]

In fact, we have many sources of knowledge, although all are without authority, and what semblance of authority exists—suspiciously fragile and doomed to continuous revision—is never beyond criticism:

> Science is not a system of certain, or well-established, statements; nor is it a system which steadily advances towards a state of finality. Our science is not knowledge (*epistēmē*): it can never claim to have attained truth, or even a substitute for it, such as probability.[4]

And what about truth? Again, we have no access, although the scientific method allows for recognition of falsity and error, and, by inference, leads us to take positions perhaps closer to "truth." Popper states, "The most important function of observation and reasoning, and even of intuition and imagination, is to help us in the critical examination of those bold conjectures which are the means by which we probe into the unknown."[5]

Nothing of the natural world is demonstrably true. Useful theories are those set forth in a manner that permits concepts of nature to be tested for any *falseness* of content. Science is the business of devising and testing theories, not establishing truths, which are metaphysical. Experiments correctly conceived and executed can therefore *refute* theories, but never confirm them. The only valid test of a theory is how well it withstands repeated attempts to falsify it (table 4).

A *scientific theory* is a statement about some aspect of the natural world, phrased so as to be amenable to falsification. Theories, in other words, have certain properties (table 4). Popper proposed three criteria for a new theory: (1) that it "proceed from some simple, new, and powerful, unifying idea about some connection or relation (such as gravitational attraction) between hitherto unconnected things; (planets and apples)"; (2) that it "be independently testable"; and (3) that it pass some "new, and severe, tests."[6]

Theories that are too broad to be tested directly can be assessed using hypotheses. *Hypotheses* are simple statements conceived to address different facets of a theory. Later, another aspect can be tested by devising a different hypothesis, and then another, and so on. Think of hypotheses as spears thrust against a theory's hide to refute its durability. Each spear-thrust is an episodic attempt at falsification. If after numerous thrusts the theory remains unpunctured, scientists might consider it to have been *corroborated*. Even after many years of such testing, the best theories are never verified.[7] Verification is impossible; again, theories can only be falsified. Proof is not a result of scientific empiricism ("scientific proof" is oxymoronic), nor is conclusive *disproof* possible either. The survival of theories in no way attests to their irrefutability.

According to Popper's model, a theory that is absolutely true is untestable and thus immune to falsification. Obviously, an irrefutable theory is not necessarily true. This can be addressed logically by considering that two irrefutable and incompatible theories are able to coexist. If their incompatibility is genuine, one must be false, although we have no way of determining which of them it is. In addition, every logical statement and its negation must both be logically unfalsifiable. Neither the statement *Today is Monday* nor its

Table 4. Thirteen properties of scientific theories.

---

1) Verification (i.e., confirmation) of a theory is easily obtained if sought, but the task of science is to refute theories, not verify them.
2) A theory can never be truly verified, although corroboration can result only if a theory withstands rigorous attempts at falsification.
3) Every good theory is a prohibition by forbidding certain things to happen; the more elements forbidden, the better the theory.
4) Nonrefutable theories are unscientific and therefore metaphysical.
5) The only proper test of a theory is the attempt to falsify it.
6) Corroboration of a theory is unacceptable unless resulting from a genuine test of falsification.
7) Once falsified, a theory cannot be resurrected by adding auxiliary, or *ad hoc*, statements, which make it more difficult to falsify in the future.
8) No theory can be deduced from statements of observation.
9) No predictions derived from a theory can be extrapolated to a universal statement (i.e., to anything beyond our experience).
10) Science does not proceed from observation to theory, but from theory to observation; we attempt to explain the known by using the unknown.
11) Most theories are too general to be tested directly, requiring hypotheses designed to test them piecemeal.
12) The simplest theories are the most testable.
13) All theoretical knowledge—and thus every theory—is uncertain; with the help of observation we can know only singular experiences. Theories consist always of opinions (*doxa*), not certain knowledge (*epistēmē*).

Compiled mostly from information in Popper 1968.

negation, *Today is not Monday,* can be refuted. Thus it follows that false statements exist that are logically irrefutable.[8] In the case of empirically irrefutable statements (i.e., those requiring observation), the situation is different, although the result is the same. Consider the statement, *There exists a pink manatee that if glimpsed at sunset on the eighteenth day of the tenth month of an uneven year bestows immortality on the viewer.* The statement is irrefutable without a means of testing it. Most would reject it as false, but the possibility of truth still remains. In testing theories we retain those that bring us closer to facts we understand; in other words, that correspond most closely with observations and predictions. Can truth ever be caught in a bottle? Never, but in seeking truth we sometimes get to smell it on the night air or glimpse it moving swiftly through the trees. Instead of crowding closer to the edge of conventional interpretation, a good hypothesis, like the good theory it tests, should be a step into the void.

Belief for the practicing scientist is never justified by observation because this would imply truth through verification. Many scientists are also (unfortunately) verificationists. They put forward the premise that any belief ought to be based on positive evidence; on "truth," in other words. At the very least, they claim, it should be probable in statistical terms and not merely possible.[9] However, the correlation of the probability of a hypothesis with the probability of events is very weak. In terms of logical form, probability is neither verifiable nor falsifiable.

Popper discounted the very notion of probability.[10] To him, "Logical probability . . . represents the idea of approaching logical certainty, or tautological truth, through a gradual diminution of informative content." I agree that as statistical noise increases, information's signal weakens. I doubt, however, that competent scientists think of statistical probability as approaching logical certainty. Scientists employ probability not as a means of quantifying truth but as a technique for hypothesis testing. If truth and falsity are absolute in logic, can their disparity be diluted in empiricism? Of course. Despite this liability, scientists still seek "truth" while recognizing the impossibility of ever finding it. Empiricism can never match formal logic in the certainty of its statements; consequently, the quality of a theory bears directly on the quality of answers derived from it.

Popper advocates *degrees* of truth on the basis of logical validity.[11] He uses as an example, if A is always true (i.e., all logical components of A), then the class A can consist only of true statements. If A is false, its content will consist of both true and false statements. *It always rains on Sundays* is false, although the conclusion that it rained last Sunday might be true. "Thus whether a statement is true or false, there may be more truth, or less truth, in what it says, according to whether its content consists of a greater or lesser number of true statements."[12]

If what makes a theory testable is its degree of vulnerability to refutation, or falsifiability, then contrary to what common sense tells us stronger theories are easier to falsify than weaker ones. The perfect metaphysical theory is unfalsifiable. The strongest scientific theories are easily falsifiable but have survived repeated tests and all the criticism science can muster.

Theories purporting to show that proof is all around are invariably weak enough to be labeled as metaphysical and therefore unscientific. Consider weather prediction. The statement, *It might rain tomor-*

row is unfalsifiable. In taking no risk the weather forecaster's prediction is above falsification: the prediction is irrefutable with or without rain. *It will not rain tomorrow* is a much stronger statement: refutation arrives with the first raindrop. These are poor examples of how science works, and good science obviously begins with asking proper questions. They make the point, however, that good theories take risks in the form of strong statements made vulnerable to falsification. The more daring a theory's content, the better it is. Science is served most efficiently if poorly conceived theories fall quickly to the wayside.

Because the ideal of *epistēmē* will never be realized and there can be no absolute, demonstrable knowledge, every scientific theory will always remain a theory. Consequently, every scientific statement, whether universal or basic, will always be tentative. The corroboration of a theory is uncertain, being susceptible to the meaning and interpretation of other corroborated statements.[13] Creationists are not faced with this fundamental dilemma because their beliefs are metaphysical, taking the form of dogma instead of refutable theories. As Popper says, "Only in our subjective experiences of conviction, in our subjective faith, can we be 'absolutely certain.'"[14] Some might ask, then what is the point of science? Put simply, science seeks answers to bold questions about Nature while recognizing that any such answers are never final. The challenge is in the pursuit of problems that lead us deeper into the rich realm of abstraction, ever closer to the core of what propels the universe.

If the notion that science seeks to verify, or prove, its theories remains its prevalent myth, the myth of induction is another that follows closely. *Induction* encompasses the belief that science begins with observation, proceeds to generalization (or hypothesis), and culminates in theory. This is fallacious. By the use of induction any theoretical statement can be constructed out of accepted basic statements, as can any metaphysical system.[15] As Rupert Riedl writes, "Logic can only transmit the truth which it possesses but cannot extend it."[16] To phrase this differently, because only simple statements exist within the realm of our experiences, the transition from simple to universal statements is impossible. We see in the epigraph the dilemma this causes Bouvard and Pécuchet, Flaubert's bourgeois fools. Induction must inevitably involve either infinite regress or the *a priori* acceptance of some synthetic principle as valid.[17] Despite a few dissenting opinions,[18] induction is ordinarily not considered part of the scientific method.

An *inductive inference* is one that advances from singular or basic *particular* statements (e.g., the results of observations or experiments) to universal statements such as theories.[19] Every scientist—indeed, almost every person—is familiar with induction. Most beliefs and many of our daily decisions are based on it, despite its falseness. In the classic example an observer notices that all the swans he sees are white. After looking at thousands of white swans he makes the universal statement, *All swans are white*. However, no matter how many white swans he sees his conclusion is unjustified. In some remote part of the world might exist a black swan. In fact, *Cygnus atratus*, a black species, was discovered in 1697 in the marshes of northeastern Australia. Even had it never existed, induction could not have been justified.

In other words, we base inductive statements on what we perceive to be the general uniformity of Nature known from experience. However, inferring that such uniformity will persist into the future extends the observed to the unobserved, or the inference for which we seek justification.[20] The result is a circular argument.

Nothing can be observed in a universal statement. The statements *All manatees are aquatic* and *All manatees are terrestrial* are not even contradictory. By being inclusive they simply imply the nonexistence of manatees. The point is, no singular statements can be deduced from them, none, that is, that can be falsified. In chapter 5 I discussed universal statements and the problems of using them in zoo graphics. Statements containing the words *there is (are)* are existential. Paraphrasing Popper, *There is one seahorse in trouble* is an *existential statement*.[21] The negation of a (strictly) universal statement is equivalent to a (strictly) existential statement, and the reverse is also true.[22] To state, *not all seahorses are in trouble* is no different from, *there is a seahorse in trouble*.

All inductive statements are predicated on some sort of experience traceable ultimately to a simple statement. However, universal truths are never reducible to simple statements, and therein lies the fallacy. A single good deed no more proves humanity's generosity than an act of atrocity is evidence of our universal cruelty. In science, the empirical process takes place in sequential steps involving *deduction*. Induction is neither evident nor necessary.[23]

We experience inductive reasoning daily, and each incident is fallacious by falling outside the experiences of everyone; that is, extrapolation from simple statement (what we can experience) to universal statement (what we can *never* experience). Economists might agree that last Wednesday's strong performance by U.S. computer stocks is evidence of Asia's rebounding economy. Stock analysts see in the narrowing sector of high-performance stocks an impending "correction," or decline in the overall market. These experts never make specific predictions that might be falsifiable. Their comments are actually guesses couched in metaphysical terms, rendering them useless and unscientific. On any given day the sales of computers will go up or down, and the Asian economy will rise or fall. Similarly, the stock market is certain to decline (it always does) or rise (it always has). Nothing about Wall Street is remotely scientific.

I discuss in chapter 2 the distinction between class (universal) names and individual (proper) names. Its importance carries over to theories, which begin as classes but are tested as individuals. The inference, in other words, is from universal (theories) to singular (predictions derived from hypotheses). Individual names or concepts must occur in all singular statements.[24] For one thing, the prediction must be restricted spatio-temporally. Instead of referring to *those large floating objects,* we must state, *West Indian manatees.* On occasion, ostensive gestures can be substituted—for example, *West Indian manatees and closely allied species in the Western Hemisphere.* However, there can be no doubt that in doing so we are still referring to individuals and not universals, or classes (i.e., "manatees").

The common mistake of confusing class and individual terms has its parallel in the fallacy of induction. Individual concepts are often used to refer to class concepts, which is the same as moving from singular to universal statements. An important difference occurs when a restricted statement contains a class name but in being restricted is not universal. *All manatees are aquatic* is a strictly universal statement. *Some manatees are aquatic* is not, although the term *manatee* in both statements refers to the class of manatees and not to individuals.[25]

Popper suggests that theories be posed as negations of existential statements. By prohibiting something, a theory becomes testable. The existential statement, *there exists an albino manatee* becomes falsifiable if restated to include a prohibition and more detail: *an albino manatee does not exist in Florida's Inland Waterway.* Theories with a high probability of resisting falsification are the weakest and least interesting. Consider the biological species definition, which states, *A species is a reproductive community of populations (reproductively isolated from others) that occupies a specific niche in nature.* Notice that nothing specifically is prohibited. Notice also that key words in the definition—

community, populations, niche—are not in themselves prohibitive, nor are their definitions any less metaphysical than *species,* the term they are intended to define. Not surprisingly, the biological species concept remains metaphysical and unscientific.

Ad hoc qualifiers insulate hypothetical statements from refutation. Suppose the prediction is made that it will not rain tomorrow, but rain is falling when morning arrives. The forecaster, hoping to salvage his credibility, announces that the day would have been sunny were it not for the arrival of an unforeseen low-pressure front and a tendency for more rainfall at this time of year. The original forecast was stated simply and directly: it prohibited rain the following day and thus was falsifiable. With the addition of two *ad hoc* statements the original prediction has been weakened beyond falsification. No longer is the simple observation of rain an adequate test. We must now include the possibility of unforeseen weather fronts and factor in the time of year. Interesting and powerful statements have the greatest potential explanatory power and the lowest probability of surviving tests of falsifiability.

Theories are science's myths, our dreams about the natural world. Theories reduce the known to the unknown, not vice versa. They *precede* observation, which seems illogical, but consider that although observation and theory are closely linked, observation has no relevance except when enveloped within a framework of theory. This is why recording observations on a pregnant aardvark fails as science. What can be observed is strictly that—an observation.[26] Facts accumulated in the absence of theory never lead to enlightenment.

An observation coupled with existing theory and posed as a problem is different. Others sitting under apple trees had undoubtedly been struck by falling apples prior to Isaac Newton's revelatory experience, but only Newton placed the event within the context of earlier theories of moving bodies. Without prior awareness that his predecessors had considered the same problem, being hit on the head by an apple would have been a forgettable event. For Newton, the search had been on *before* sitting down underneath a tree; otherwise, the connection of an accident with imposition of a natural law could never have been made. As Heraclitus wrote, "He who does not expect the unexpected will not detect it: For him it will remain undetectable, and unapproachable."[27]

Zoos claim falsely to *conserve* biodiversity when what they really do is *preserve* isolated fragments of the biosphere, either as phenotypes in cages, paddocks, and aquariums or as genotypes frozen in liquid nitrogen. As artifacts, semiotically real animals and their frozen alleles both retain a certain material presence, although nothing that connects them directly with Nature. To quote O. H. Frankel and Michael E. Soulé:

> We use the term "conservation" to denote policies and programmes for the long-term retention of natural communities under conditions which provide the potential for continuing evolution, as against "preservation" which provides for the maintenance of individuals or groups but not for their evolutionary change. Thus, we would state that zoos and gardens may preserve, but only nature reserves can conserve.[28]

Largely as a result of this situation zoo-based science is denied suitable hypotheses needed to test biology's important theories, in particular evolution in its many guises. Natural selection, distribution and biogeography, the nature and functioning of ecosystems—all are well beyond any zoo's horizon. The available theories are trivial in comparison and shadowed by the chronic anamensis of restricted space. Needless to say, hypotheses bounded entirely by the artifactual are divorced from Nature, and extrapolating experimental results beyond captivity is a parlous reach, except perhaps for inductionists.

Every proselytism about saving species has the moldy odor of modernism. Especially dated is treatment of the terms *species* and *biodiversity* as synonyms when in fact they have little in common. Brent D. Mishler is one of a growing number of evolutionary biologists who advocates eliminating the species concept entirely. Like all ranks in the Linnaean hierarchy the species rank is arbitrary, containing no greater measure of reality than a photographic image. Moreover, as constructs of the human imagination "species are not comparable in any important sense and cannot be made so."[29] The loss of a parasite dependent on a single host scarcely compares with the loss of an army ant that hosts at least one hundred other species from mites to antbirds.[30] Mishler argues correctly that making conservation decisions based on named species is "mindless," the cataloguing of species "misguided." There are better ways to assess and conserve biodiversity, but zoos have not yet heard the message.

Trapped inside its restricted sphere, zoo biology proceeds as it always has, devising experiments and publishing results that satisfy contemporary peer *doxa* but seem irrelevant in the larger arena. As Elgin writes, science "wants to discover the number of fundamental physical forces in nature, not the number of lamp posts in Harvard

Yard."[31] If zoos indeed become arks without a place to set their cargos ashore then the effort will have been entirely self-serving.[32] And if this comes to pass the future zoo will be just what it is today—a modernist collocation with the seventeenth century menagerie at Versailles, a place where spectators can gawk at beautiful and rare creatures, any debate over an epistemic link with Nature having been long forgotten.

Notes
Introduction

1. Baudrillard 1994, 75.
2. Ibid., 82.
3. Benjamin 1969, 159.
4. Friedberg 1993, 1.
5. Jameson 1983, 125.
6. Benjamin 1978, 249.
7. Pamuk 2004, 29.
8. Friedberg 1993, 7.
9. Ibid., 3.
10. Loomis 2003.
11. Benedikt 1994, 2.
12. Jameson 1983, 125.
13. Ibid., 114.
14. Benjamin 1978, 158.
15. Link 1883, 1226.
16. Baudrillard 1994, 1, 45.
17. Ibid., 2.
18. Ibid., 30.
19. Ibid.
20. When modernism began and ended is arbitrary, but most cultural historians give its life span as 1860 through 1959 with postmodernism beginning with the 1960s. Lyotard (1991, 25) writes, "the result is that neither modernity nor so-called postmodernity can be identified and defined as clearly circumscribed historical entities, of which the latter would always come 'after' the former."
21. Histories of zoos and aquariums, of course, exist. See, for example, Hanson (2002) and Kisling (2001). Critiques of zoos are legion, most written from ethical or sociological viewpoints. I cite several in the following chapters.
22. Spotte and Clark 2004.
23. Benjamin 1978, 63.

Chapter 1.

1. Semiotics was founded independently in the early twentieth century by the Swiss linguist Ferdinand de Saussure and the American philosopher Charles Sanders Peirce. Their methods and interpretations differ in important respects, but in my application elements of both are used following Chandler (2002) and Eco (1976), although I rely more heavily on Peirce. The de Saussurean model treats (in his own words) the signifier and signified as if they were joined together permanently. How-

ever, this disregards how portions of either semiotic model dissociate and recombine in different configurations, like chemical electrolytes in a solution (see Ulmer 1983). On how semiotics describes contemporary culture, Jameson (1991, 96) writes: "Now reference and reality disappear altogether, and even meaning—the signified—is problematized. We are left with that pure and random play of signifiers that we call postmodernism, which no longer produces monumental works of the modernist type but ceaselessly reshuffles the fragments of preexistent texts, the building blocks of older cultural and social production, in some new and heightened bricolage." For a thorough treatment of Peircean semiotics see Merrell (1997). For the original treatments see Peirce (1966) and de Saussure (1983).

2. Ogden and Richards 1960, 79.

3. Peirce greatly confused the concept of the sign by also using this term to prop up one corner of his triad. In de Saussure's dyad the sign's role is much clearer by simply representing the sum of the signifier and the signified.

4. See Ogden and Richards 1960, 10–15.

5. Sebeok 2001, 11.

6. Ibid., 10.

7. Ibid.

8. See Elgin (1997, 137–146) for an interesting critique of Peirce's original usage of symbol, icon, and index and its effects.

9. Eco 1979, 179.

10. Eco 1984, 17.

11. Eco 1979, 182.

12. See Ibid., 190–92.

13. Eco 2000, 92, Merrell 1995, 76.

14. Eco 2000, 181.

15. Joyce 1961, 186.

16. These comments are not an endorsement of essentialism.

17. Eco 2000, 132.

18. This adoption of Eco's terminology to zoo graphics is mine, and I take responsibility for any misinterpretations. It seems better than inventing a neologism. See Ibid., 136–141.

19. Ibid., 139.

20. See Spotte and Clark 2004.

21. See Elgin 1983, 43–58.

22. Eco 2000, 147.

23. See the extended argument of Ghiselin (1997) for species as individuals.

24. Merrell 1995, 133.

25. Ibid., 225.

26. Nature—the Other—is distant even when nearby. Donato (1979, 231) says, "Behind our gardens and our fields hides a Nature to which we cannot have access."

27. The Other in philosophy is used to designate *other* than humankind. In theology, it refers to the concept of God, a being so remote and strange as to be incomprehensible (e.g., Pagels 2003, 181). Not surprisingly, Nature can assume mystical properties in "deep ecology" and similar branches of postmodern environmentalism.

28. Merrell 1995, 130.

29. Wilden 1987, 245, Merrell 1995, 131.

30. Cobley and Jansz 1997, 22.

CHAPTER 2.

1. Borges 1998, 325.
2. Baudrillard 1994, 1–2.
3. See Ghiselin (1997) for a thorough analysis of the metaphysics of individuality.
4. Hofstadter and Dennett 1981, 3–7.
5. Stille 2002, 41.
6. The *Morgan* is a later incarnation of the ship of Theseus, a puzzle known since antiquity. A vessel once belonging to Theseus was kept in Athens for many years. Over time, all its planks were gradually replaced until no part of the original vessel remained. Was it the same ship? Restoration and identity is of major concern in the art world (see Elgin 1997, 97–109).
7. Hull 1989, 86.
8. Eco 2000, 212.
9. Baudrillard 1983, 83.
10. Baudrillard 1994, 121. Baudrillard is never clear on the limits of these definitions in his English edition of *Simulations*. Undoubtedly, counterfeit copies of tools and articles of clothing were produced during humankind's evolution, and evidence is abundant from ancient times (e.g., Cuomo 2003).
11. According to McLuhan (1969, 153), "print was the first mass-produced 'commodity.' The assembly line of movable types made possible a product that was uniform and as repeatable as a scientific experiment."
12. Spotte 2002, 223.
13. Ibid.
14. Ibid., 224.
15. Baudrillard 1994, 16.
16. Jameson 1991, 18.
17. For example, see the hallucinatory novels and stories by the Latin American writers Jorge Luis Borges, Julio Cortázar, Gabriel García-Márquez, and Juan Rulfo. Among U.S. postmodern writers, the imaginative works of John Barth, Donald Barthelme, Robert Coover, and Barry Hannah stand out.
18. Merrell 1995, 260.
19. Lee 1999, 50.
20. Ibid.
21. Goodman 1976, 437.
22. Ibid., 438.
23. Because mucus from the skin of these frogs is used to make arrow poison—not arrows containing poison—they logically should be called "arrow-poison frogs." Most naturalists refer to them as "poison arrow frogs," which not only is illogical, it makes little sense grammatically. *Poison* is properly a noun or verb; the more acceptable adjective is *poisonous*.
24. See Donato 1979, 230.
25. Ibid., 223.
26. Ibid., 224.
27. Ibid.
28. Vidler 2000, 151.
29. Harbison 1977, 143.
30. Ibid., 131.
31. Ibid., 144–45.

32. Ibid., 140.

33. Père David's deer once ranged through the northeastern and eastern central parts of China, but it was evidently hunted to extinction in the wild more than a thousand years ago and survived only in parks. In the nineteenth century Père David, a French missionary, saw the last extant specimens in a herd belonging to the emperor. Some were subsequently shipped to Europe and bred successfully. The remaining animals in China died in the early twentieth century. In the late 1980s, specimens from the West were released into China's Dafeng Reserve where they are multiplying.

34. Hayles 1995, 410.
35. Lee 1999, 95.
36. Ibid., 53.
37. Benjamin 1978, 94.

Chapter 3.

1. Eco 1986, 43.
2. Sorkin (1992, xi) calls the ageographical city "a city without a place attached to it." Much of it exists in cyberspace and, according to Sorkin, "this city looks a lot like television."
3. Ibid., xv.
4. Crawford 1992, 3–4.
5. See www.MahmoudiehConcepts.com
6. Benjamin 1978, 86.
7. Eco 1986, 51. For the shopping mall as Eden see Merchant (1995).
8. Mullan and Marvin 1987, 6–7. On page 3 the authors write: "Once brought to human attention an animal is no longer simply an animal in itself—it can only be that away from human sight, experience and thought."
9. Ibid., 12.
10. In 1977, during my tenure as a public aquarium director, I went to the Pribilof Islands in the Bering Sea, captured 10 adult northern fur seals, and brought them back to be put on exhibit. I had given them numbers in the field, and as the weeks passed these inevitably became their names.
11. Spotte 2002, 227.
12. Eco 1986, 44.
13. Schickel 1968, 323.
14. Lee 1999, 140.
15. The source of these ideas is Ulmer (1983).
16. Cortázar 1963, 3–8.
17. Ibid., 4.
18. Ibid., 7.
19. Malamud and I read this story differently. To me, Cortázar's intention is purely artistic, and any interpretation based on a contemporary social theory is misdirected. To Malamud (1998, 247), "His [Cortázar's] point is that spectatorship, which he depicts as evil in its exploitative consumption of an objectified animal, debases the watcher as well as the watched." Interesting, but Malamud can't know Cortázar's motive. And even if he did the result would be the same. As Sontag (1961,

26) writes: "A work of art, so far as it is a work of art, cannot—whatever the artist's personal intentions—advocate anything at all."

20. Hargrove 1995, 14.

21. As Hayles (1995, 441) writes: "When 'nature' becomes an object for *visual* consumption, to be appreciated by the connoisseur's eye sweeping over an expanse of landscape, there is a good chance it has already left the realm of firsthand experience and entered the category of constructed experience that we can appropriately call simulation."

22. Eco 1986, 10.

23. Calvino 1988, 86.

24. Hargrove 1995, 15.

25. Conway 1995, 1.

26. Wieseltier 2004, 39.

27. Hargrove 1995, 17.

28. Ibid., 17–18.

29. Ibid., 18.

30. Borgmann 1995, 39.

31. Shepard 1995, 22.

32. Ibid.

33. See McKibben 1989.

34. Rothfels (2002, 212) also makes this point, giving its genesis as Hagenbeck's Animal Park: "The metaphor of the Ark earned the Park, together with almost all zoos in the twentieth century which adopted the idea, a profoundly resonant justification for their continued existence in the face of their critics." Also according to Rothfels (page 212), Hagenbeck was among the first to promote the idea that visiting his park helped to protect wild animals.

35. Ibid., 199.

36. See Hyson 2000.

37. Ibid., 25.

38. Mullan and Marvin 1987, 54.

39. Ibid., 71. See Spotte (2002, 220–31) for a fictional account of a comparable experience.

40. Baudrillard 1994, 3.

41. Nagel 1974.

42. Gorilla's amazing leap, 2004.

43. Martel 2001, 41.

44. Goodman 1968, 6–7.

45. Ibid., 5.

46. Baudrillard 1994, 6.

47. Ibid., 11.

48. Benjamin 1978, 158.

49. Ibid.

50. Hardwick 2004, 30.

51. Benjamin 1969, 240.

52. Ibid., 199.

53. Ibid.

Chapter 4.

1. Todorov 1977, 89.
2. Cohan and Shires 1988, 1.
3. My use of *spectacle* differs in important respects from how the term is used in narrative theory (e.g., Bordwell 1985).
4. R. Chandler 1976, 161.
5. Goodman 1968, 69.
6. Eagleton 1983, 166–67.
7. Ibid., 101. Recent semioticians have replaced de Saussure's term *associative* with *paradigmatic*, an unfortunate choice. First, the dictionary definition of *paradigm* refers to a conjugation or declension *of a word*, not a sequence of words. Second, since Kuhn (1970), a *paradigm* has come to mean a body of unprecedented scientific theory (e.g., Lavoisier's *Chemistry*) that serves for a time as the standard while simultaneously allowing intellectual room for further advancement and perhaps the future emergence of a different paradigm. Hassan (1987, 119–121) briefly discusses Kuhn's paradigms versus those of the humanities. For one thing, dominant paradigms do not exist in the humanities, and no critical argument can "invalidate" another. Not surprisingly, in the humanities every act is self-consciously historical; in the sciences nothing ever is.
8. D. Chandler 2002, 81.
9. Cohan and Shires 1988, 14.
10. Ibid., 12.
11. Ibid., 12–13.
12. Ibid., 13.
13. D. Chandler 2002, 81.
14. Cohan and Shires 1988, 15.
15. Here I mean truth in a linguistic context, not in the context of science (see chapter 11).
16. Hagoort et al. (2004) demonstrated that a statement, whether written or spoken, is processed in the brain's left inferior prefrontal cortex where meaning and knowledge are recruited and integrated simultaneously within four hundred milliseconds. In addition, the brain records and files away what makes a sentence difficult to interpret based on flawed semantic content, incomplete world knowledge, or both.
17. Near the end of his text on literary theory Eagleton (1983, 204) makes this surprising statement: "The final logical move in a process which began by recognizing that literature is an illusion is to recognize that literary theory is an illusion too."
18. Iser 1978, 54.
19. Goodman 1968, 68–85.
20. Elgin 1997, 12.
21. D. Chandler 2002, 84.
22. Eagleton 1983, 118.
23. Eco 1979, 109.
24. Ibid., 176.
25. Baudrillard 1994, 47.

Chapter 5.

1. Eco 1979, 7.
2. A *linguistic code* comprises the system of symbols used to communicate in a given language. The symbols selected help transmit the message, form an umbrella of social control over those who participate, and aid the recipient of the message to formulate a response. In U.S. zoos the linguistic code of graphic texts must be American, not British or Canadian or Bermudan.
3. Eco 1979, 8.
4. Hassan (1987, 115) calls *Finnegans Wake* "a monstrous prophecy of our postmodernity."
5. Cobley and Jansz 1997, 29, presumably quoting Peirce.
6. Eagleton 1983, 118.
7. Iser 1978, 64.
8. The basis of long-term memory is clearly biochemical. The consolidation of memory depends on brain-derived neurotrophic factor (BDNF) rather than on the transcription factor Zif268. Reconsolidation of memory reverses these requirements (Lee et al. 2004).
9. This statement is valid within the correct contexts of time and place. Elizabethan English and contemporary American slang, for example, are both unfamiliar to modern Britons.
10. Iser 1978, 67.
11. Ibid., 70.
12. Ibid., 116.
13. Eagleton (1983, 74) puts it this way: "Literary texts do not exist on bookshelves: they are processes of signification materialized only in the practice of reading. For literature to happen, the reader is quite as vital as the author."
14. Iser 1978, 21.
15. Ibid., 34.
16. Ibid., 22.
17. Ibid., 35.
18. Ibid., 36.
19. Ibid., 38.
20. Ibid., 46.
21. Ibid., 83.
22. Ibid., 109.
23. McLuhan 1969, 57.
24. Iser 1978, 118.
25. See Metzinger (2003) for a lucid description of these and other conscious experiences.
26. Krasner 1992.
27. Wilden 1987, 125.
28. Ghiselin 1997.
29. Flaubert 1976, 208.
30. Elgin 1997, 69.
31. Benjamin 1969, 89.
32. Ibid., 69.
33. McConnell 1979, 4.

Chapter 6.

1. Amis 1997, 127.
2. Daniel (2003) reported that New England Aquarium, in Boston, was forced to lay off twenty percent of its staff, mainly because of declining attendance. However, the previous year had set an attendance record attributed in part to the opening an Imax theater. More recently the Los Angeles Zoo installed an exhibit of robotic dinosaurs that move, powered by computer chips and compressed air (*USA Today*, 2004).
3. The source of this statement is always cited as Debord (1977), but it isn't there. I therefore don't know the source.
4. Busch Gardens, in Tampa, Florida, replaced its dolphin show with a 4D movie theater. Jacobs (2002, 1D-2D) writes: "The theater will . . . feature a special 3-D film called 'R. L. Stine's Haunted Lighthouse,' by the author of Goosebumps children's horror books. The theater's technology is called 4-D: a 3-dimensional movie complete with glasses and extra effects such as misting rain or rumbling seats. . . . 'It's going to be a fun and spooky adventure,' said Gerard Hoeppner, spokesman for Busch Gardens. 'The [dolphin] show was popular with guests, but they're looking for new types of thrills, and we think that will fit that bill.'"
5. Vidler 2000, 100.
6. Bousé 2000, xiv.
7. Andrew 1976, 89.
8. Bousé 2000, 6–7.
9. Ibid., 5.
10. Metz 1982, 44.
11. Ibid., 139.
12. O'Brien 1993, 16.
13. Bordwell 1985, 3. As I mentioned in an endnote in chapter 4, my use of *spectacle* differs from how the term is used in literary and film theory for which Bordwell, in recounting the history of diegesis and mimesis, gives some examples.
14. Metz (1974, xv) defines *diegesis* as "the denotative material of a film." He notes that its complement is not found in either linguistics or semiotics.
15. Bordwell 1985, 9.
16. In describing a "naturalistic" exhibit, Mullan and Marvin (1987, 78) tell us: "Although it is probably as close as the designers can come to the replication of the natural world, it is in fact a stage-set, not a real functioning ecosystem."
17. Ibid.
18. Bazin 1971, 95.
19. Eco 1986, 152.
20. Bazin 1971, 104.
21. Ibid.
22. Ibid.
23. See Benjamin 1969, 191.
24. Bordwell 1985, 9.
25. Ibid.
26. Ibid., 10.
27. For examples of these early works on film theory see Arnheim (1957), Balázs (1970), Bazin (1971), Eisenstein (1957a, 1957b), and Kracauer (1965). Later works

include Bordwell (1985, 1989), Easthope (1993), Lapsley and Westlake (1988), and Monaco (1981).

28. Linden 1970, 41.
29. See Barthes 1967, 1972, 1983.
30. Anyone can identify "visual language" construction in films, but homology is another matter. An example of "simile" can be found in *Lawrence of Arabia* when Peter O'Toole blows out a flame and a desert scene under a blazing sun explodes across the screen.
31. Metz (1974, 67) argues, "The image is always speech, never a unit of language." The syntax of language has no analogue in film. I agree. Paradigmatic units are large and usually obvious (e.g., bad guys in black hats, good guys in white hats).
32. Bazin (1971, 114) writes: "A film calls for a certain effort on my part so that I may understand and enjoy it, but it does not depend on me for its existence."
33. Linden 1970, 2, 4.
34. Metz 1974, 10.
35. Linden 1970, 19.
36. Ibid., 32. The phrase *negates itself* is one of those unsatisfying and opaque images favored by theorists of the arts, unsatisfying in being an empty sign, opaque in lacking literal meaning. *Steps outside us* is clearer metaphor.
37. Jameson 1991, 75.
38. See Iser 1978, 36–37.
39. See Spotte 1992, 555–56.
40. Metz 1974, 23.
41. Ibid., 45–46. When friends meet and one tells the other about a film she has just seen, plot is usually mentioned before anything else (i.e., *It's about these two people who meet in Paris during the War, except one of them is secretly a spy.*).
42. Metz 1974, 46. I would rephrase this as going from no narrative to narrative.
43. Ibid., 5.
44. Ibid.
45. Ibid., 22.
46. Andrew 1976, 202.
47. Metz 1974, 7.
48. Burch 1973, 40.

Chapter 7.

1. Bordwell 1985, 30–47.
2. Bordwell's use of *schemata* in a decision-making context seems similar to Kant's (chapter 1).
3. Bordwell 1985, 31.
4. Ibid., 32.
5. Linden 1970, 40.
6. Ibid., 88.
7. Vidler 2000, 158.
8. Hasson et al. 2004, Pessoa 2004. The visual stimulus used by Hasson and coauthors was an uninterrupted thirty-minute segment of *The Good, the Bad, and the Ugly*, released in 1966 and starring Clint Eastwood.

9. See Cox et al. (2004), who also used fMRI in their study of the human neurological response to contextual stimuli.

10. Bordwell 1985, 33.

11. McConnell 1979, 5.

12. Bordwell 1985, 35.

13. Linden 1970, 63.

14. Friedberg 1993, 15.

15. Ibid., 4.

16. Ibid., 5. Even postmodern malls pay tribute to modernist Parisian shoppers. Crawford (1992, 3) describes the West Edmonton Mall as containing some wings that "mimic nineteenth-century Parisian boulevards."

17. Friedberg 1993, 4.

18. See Ibid., 30, for the source of this passage.

19. Ibid., 65.

20. Ibid., 66.

21. See Ibid., 66 and 68, for references in support of this analogy.

22. My account of Robertson's performance is mainly from Barnouw (1981, 19–24), but also see Christopher (1962, 137).

23. Barnouw 1981, 19.

24. Christopher (1962, 137), who makes no mention of newspapers, claims Robertson used blood, vitriol, and alcohol. However, Robertson (1831, 131) writes that he threw on the fires "*deux verres de sang, une bouteille de vitriol, douze gouttes d'eau-forte, et deux exemplaires du journal des Hommes-Libres.*" I translate this as, "two glasses of blood, a bottle of vitriol, twelve drops of aqua fortis, and two copies of the newspaper *Free-Men*." Aqua fortis is another name for nitric acid. I doubt if this procedure would meet OSHA regulations if used today.

25. Barnouw 1981, 19.

26. Marion, in *L'Optique* (1869, 212), describes the process somewhat differently: *Le rideau de percale, d'au moins 20 pieds carrés . . . sera enduit d'un vernis composé d'amidon blanc et de gomme arabique choisie, afin de le rendre légèrement diaphane.* [The curtain of percale, of at least twenty square feet . . . will be smeared with a varnish composed of white starch and gum arabic, chosen to make it slightly diaphanous.]

27. *il s'efforçait au contraire d'établir, aux yeux de tous, l'absence de tout cause occulte et l'action seule de procédés scientifiques.* Marion 1869, 213.

28. Gernsheim and Gernsheim 1956, 5.

29. Was this the artist Jacques-Louis David? See Benjamin 1978, 149.

30. See Gernsheim and Gernsheim 1956, 14.

31. Ibid., 15.

32. Ibid., 18.

33. Sternberger 1977, 13.

34. Friedberg 1993, 25.

35. The source of this information is Lorenz Hagenbeck (1956, 27–28, 36, 77, 98–99). Lorenz was Carl's son, and his book, written late in life, is a memoir. The design is never fully described. Moats were later criticized by Hediger (1969, 192–93), who referred to them as "ditches" and considered their improper design and use as dangerous to both beast and humankind.

36. Burch 1973, 38.

37. Ibid., 35.

38. According to Malamud (1998, 230), Michel Foucault infers that the design of La Vaux's menagerie at Versailles "exerted an institutional influence upon Bentham." Mullan and Marvin (1987, xviii, 31–45, 101–3) give a historical overview of the similarities among prisons, insane asylums, and zoos.

39. Friedberg 1993, 19.

40. Ibid., 17.

41. It seems strange that the panopticon's inventor is today considered a hero among so-called "animal rightists." In the hands of admirers like Peter Singer, Bentham's utilitarian philosophy extends to animals the rights and freedoms never envisioned by Bentham, least of all to human beings occupying his hypothetical panopticon. Mainly zoos have adopted Bentham's design, where once again its purpose is confinement and the inmates are subjected to the keeper's and the spectator's mobile (and presumably) perpetual gaze.

42. Friedberg 1993, 20.

43. Among the earliest were guided tours of continental Europe offered by Thomas Cook of London in 1855.

44. Friedberg 1993, 61.

45. Ibid., 59.

46. Fisher 2002, 32–33. Serrell (1998, 9) mentions the similarity between shopping and viewing museum exhibits.

47. Friedberg 1993, 53.

48. See Spotte 1975.

49. Eco 1986, 294.

50. Benjamin 1978, 155.

51. Crawford 1992, 13.

52. Ibid., 14.

53. Hardwick 2004, 4.

54. Ibid., 148.

Chapter 8.

1. Arnheim 1974, 17.

2. I use *photography* and *photographs* to denote still photography and *cinema, film,* or *the movies* to denote moving pictures.

3. "Starting from a few personal impulses, I would try to formulate the fundamental feature, the universal without which there would be no Photography." Barthes 1981, 8–9.

4. Sontag 1973, 85. Also see Spotte 2002, 220–31.

5. Sontag 1973, 5.

6. Hull 1989, 95.

7. Sontag 1973, 80.

8. Barthes 1981, 5.

9. Baudrillard 1994, 117.

10. Goodman and Elgin 1988, 111.

11. Goodman 1968, 5, 21–26.

12. Barthes 1981, 80.

13. Ibid., 34.

14. Sontag 1973, 110.

NOTES

15. Barthes (1981, 85) wrote: "The Photograph does not necessarily say *what is no longer*, but only and for certain *what has been*." Also see Metz 1974, 6–7.
16. Ibid., 78.
17. Ibid., 98.
18. Ibid., 9.
19. Ibid., 92.
20. Ibid., 78.
21. Ibid.
22. Metz 1974, 46. Famous photographs (e.g., the U.S. Marines planting the flag on Iwo Jima, Che Guevara's body displayed in a barracks) convey a story because we already know it. In this sense photographs augment narrative.
23. Goodman and Elgin 1988, 110.
24. Ibid.
25. Sontag 1973, 111.
26. Ibid., 145, footnote.
27. Ibid., 120.
28. Jameson 1983, 125.
29. Ibid., 120.
30. Debord 1977, #18.
31. Sontag 1973, 165.
32. Ibid., 97–98.
33. Ibid., 7.
34. Ibid., 4.
35. Ibid., 14–15.
36. Barthes 1981, 91.
37. Ibid., 27.
38. Ibid., 51.
39. Ibid.
40. Ibid., 55.
41. After photographing this scene I caught the seal and cut away the net.
42. Barthes 1981, 12.
43. Ibid., 81.
44. Sontag 1973, 15.

CHAPTER 9.

1. Sontag 1973, 144. Sontag states on page 139: "Indeed, the difference between a good photograph and a bad photograph is not at all like the difference between a good and a bad painting. The norms of aesthetic evaluation worked out for painting depend on criteria of authenticity (and fakeness), and of craftsmanship—criteria that are more permissive or simply non-existent for photography. And while the tasks of connoisseurship in painting invariably presume the organic relation of a painting to an individual body of work with its own integrity, and to schools and iconographical traditions, in photography a large individual body of work does not necessarily have an inner stylistic coherence, and an individual photographer's relation to schools of photography is a much more superficial affair."
2. Goodman 1968, 115.
3. Ibid., 112.

4. Brinkley 2004, 264.
5. Andrew 1976, 80.
6. Sontag 1973, 92.
7. Schama 1995, 12.
8. Crimp 1983, 51.
9. Ibid., 53.
10. Benjamin 1969, 217–51.
11. Sontag 1973, 154.
12. Ibid., 158.
13. Ibid., 147.
14. Ibid., 52.
15. Ibid., 58.
16. Ibid., 79.
17. Muggeridge 1974, 226.
18. Sontag 1973, 23.
19. See Elgin 1997, 64.
20. Sontag 1973, 112.
21. Ibid., 136.
22. Barthes 1981, 38.
23. Ibid., 41.
24. Sontag 1973, 97.
25. Ibid.
26. Barthes 1981, 47.
27. Metz 1974, 9.

Chapter 10.

1. Abbott 1952, 31.
2. Our eyes contain two-dimensional arrays of photon-sensing cells that detect reflected light. What we see—and subsequently perceive—are stereoscopic images in two dimensions. The mind provides the third dimension.
3. Kepes 1965, iii.
4. Pun intended.
5. See the harrowing story of Elaine Bartlett (Gonnerman 2004). The case is made that generations of poor Americans know only a series of prison cells where birthdays and holidays are celebrated in the visiting room. Confinement becomes a normal way of life. One generation follows its predecessor like captive-raised animals returning to the cage voluntarily because there is nowhere else to go, at least nowhere that feels familiar. We seem to breed both animals and human beings for a life behind bars.
6. Sontag 2004, 6.
7. Vidler 2000, 153.
8. Ibid.
9. Ibid., 238.
10. Hayward and Franklin 1974, Spreiregen 1965, 75.
11. Hayward and Franklin 1974, 45. The authors note that the H/D ratio conforms to the size-distance invariance hypothesis (SDIH) used in psychology: $\theta = S/D$

where θ is the retinal angle subtended by a perceived object, S, and D is the perceived distance of the object.

12. Spreiregen 1965, 75.
13. Ibid.
14. Hayward and Franklin 1974, 51.
15. Large carnivores do especially poorly in confinement (Clubb and Mason 2003).
16. Bamford 2000.
17. Benedikt 1979, 47, abstract.
18. Batty 2001, 123.
19. Batty (2001) describes the formal equations and few elementary statistics needed for isovist analysis.
20. Benedikt 1979, 48.
21. Vidler 2000, 246.
22. Loomis (2003, 27) writes that "it is indeed an enormous intellectual challenge to appreciate that the very three-dimensional world in which we normally act is an elaborate perceptual representation." The physical world exists to others, but is independent of the observer (Hershenson 1999). Still, the existence of an external reality must be inferred.
23. Hayles 1995, 413.
24. Errors of scale are common in young children, who attempt to sit on doll furniture or open the doors of toy cars to get in (DeLoache et al. 2004).
25. Our minds perform a similar function when reading. This appeared in 2004 on the Internet: "Aoccdrnig to a rscheearch at Cmabrigde Uinervtisy, it deosn't mttaer in waht oredr the ltteers in a wrod are, the olny iprmoetnt tihng is taht the frist and lsat ltteer be at the rghit pclae. The rset can be a total mses and you can sitll raed it wouthit porbelm. Tihs is bcuseae the huamn mnid deos not raed ervey lteter by istlef, but the wrod as a wlohe."
26. See Kennedy 2001.
27. See Arnheim 1974, 20–23.
28. Ibid., 20.
29. Ibid.
30. For this and other factors affecting weight see Ibid., 24–25.
31. Ibid., 45.

Chapter 11.

1. Popper 1968, 21–24.
2. Ibid., 25.
3. Eagleton 1983, 62.
4. Popper 1965, 278.
5. Ibid., 28.
6. Ibid., 241–42.
7. In April 2004, NASA launched a $700 million satellite designed to test another facet of Einstein's theory of general relativity (Seife 2004).
8. Popper 1965, 195.
9. Ibid., 278.

10. In neither of Popper's books cited here did he provide an unambiguous definition of "probability."
11. Popper 1968, 232.
12. Ibid., 233.
13. Popper 1965, 280.
14. Ibid.
15. Ibid., 266.
16. Riedl 1984, 144.
17. Popper 1968, 289. A *synthetic statement* is an assertion about reality.
18. See, for example, Alfred (1987) and Hanson and Bloom (1999).
19. Popper 1965, 27. For a current redress of Popper's philosophy see Lipton (2005).
20. See Salmon 1995, 651.
21. Popper 1965, 68.
22. *Some* seahorses are indeed "in trouble" (e.g., Teske et al. 2003).
23. For a refutation from the standpoint of symbolic logic see Popper and Miller (1983).
24. Popper 1965, 64
25. Ibid., 68.
26. Popper 1968, 128. Todorov (1977, 31) is correct when he says "but description itself is not science and becomes so only when it is integrated within a general theory."
27. Popper 1968, 147.
28. Frankel and Soulé 1981, 4.
29. Mishler 1999, 313.
30. This army ant is *Eciton burchelli*. See Koh et al. 2004.
31. Elgin 1997, 19.
32. Although the returns are barely in, there is evidence that extending captivity for many generations reduces the capacity to survive in the wild, diminishing evolutionary fitness (McPhee 2003). Thus the progeny of lengthy captive breeding programs might perish even if released into a world restored to Edenic perfection.

Literature Cited

Abbott, E. A. 1952. *Flatland: A romance of many dimensions.* New York: Dover Publications.

Alfred, B. M. 1987. *Elements of statistics for the life and social sciences.* New York: Springer-Verlag.

Amis, M. 1997. *Night train.* New York: Vintage International.

Andrew, J. D. 1976. *The major film theories: An introduction.* Oxford: Oxford University Press.

Arnheim, R. 1957. *Film as art.* Berkeley: University of California Press.

———. 1974. *Art and visual perception: A psychology of the creative eye,* The new version. Berkeley: University of California Press.

Balázs, B. 1970. *Theory of the film: Character and growth of a new art.* New York: Dover Publications.

Bamford, C. L. 2000. The perception of volumetric form. Ph.D. Diss., Arizona State University.

Barnouw, E. 1981. *The Magician and the cinema.* New York: Oxford University Press.

Barthelme, D. 1982. The falling dog. In *Sixty stories.* New York: E. P. Dutton.

Barthes, R. 1967. *Elements of semiology.* New York: Hill and Wang.

———. 1972. *Mythologies.* New York: Noonday Press.

———. 1981. *Camera lucida: Reflections on photography.* New York: Noonday Press.

———. 1983. *The fashion system.* New York: Hill and Wang.

Batty, M. 2001. Exploring isovist fields: Space and shape in architectural and urban morphology. *Environment and Planning* 28B:123–50.

Baudrillard, J. 1983. *Simulations.* New York: Semiotext(e).

———. 1994. *Simulacra and simulation.* Ann Arbor: University of Michigan Press.

Bazin, A. 1971. *What is cinema?* Berkeley: University of California Press.

Benedikt, M. L. 1979. To take hold of space: Isovists and isovist fields. *Environment and Planning* 6B:47–65.

———. 1994. Physics for phantoms. Columbus, OH: Softworlds Inc., Wexner Center for the Arts.

Benjamin, W. 1969. *Illuminations.* New York: Shocken Books.

———. 1978. *Reflections: Essays, aphorisms, autobiographical writings.* New York: Harcourt Brace Jovanovich.

Bordwell, D. 1985. *Narrative in the fiction film.* Madison: University of Wisconsin Press.

Bordwell, D. 1989. *Making meaning: Inference and rhetoric in the interpretation of cinema.* Cambridge, MA: Harvard University Press.

Borges, J. L. 1998. On exactitude in science. In *Jorge Luis Borges: Collected fictions*. New York: Penguin Books.

Borgmann, A. 1995. The nature of reality and the reality of nature. In *Reinventing nature? Responses to postmodern deconstruction*, eds. M. E. Soulé and G. Lease. Washington, DC: Island Press.

Bousé, D. 2000. *Wildlife films*. Philadelphia: University of Pennsylvania Press.

Brinkley, D. 2004. *Windblown world: The journals of Jack Kerouac 1947–1954*. New York: Viking.

Burch, N. 1973. *Theory of film practice*. New York: Praeger Publishers.

Calvino, I. 1988. Note. In *Under the jaguar sun*. San Diego: Harcourt Brace Jovanovich.

Chandler, D. 2002. *Semiotics: The basics*. London: Routledge.

Chandler, R. 1976. *The lady in the lake*. New York: Vintage Books.

Christopher, M. 1962. *Panorama of magic*. New York: Dover Publications.

Clubb, R., and G. Mason. 2003. Captivity effects on wide-ranging carnivores. *Nature* 425:473–74.

Cobley, P., and L. Jansz. 1997. *Introducing semiotics*. Cambridge, UK: Icon Books.

Cohan, S., and L. M. Shires. 1988. *Telling stories: A theoretical analysis of narrative fiction*. New York: Routledge.

Conway, W. 1995. Zoo conservation and ethical paradoxes. In *Ethics of the ark: Zoos, animal welfare, and wildlife conservation*, eds. B. G. Norton, M. Hutchins, E. Stevens, and T. L. Maple. Washington, DC: Smithsonian Institution Press.

Coover, R. 2002. *The adventures of Lucky Pierre: Directors' cut*. New York: Grove Press.

Cortázar, J. 1963. Axolotl. In *Blow-up and other stories*. New York: Collier.

Cox, D., E. Meyers, and P. Sinha. 2004. Contextually evoked object-specific responses in human visual cortex. *Science* 304:115–17.

Crawford, M. 1992. The world in a shopping mall. In *Variations on a theme park: The new American city and the end of public space*, ed. M. Sorkin. New York: Noonday Press.

Crimp, D. 1983. On the museum's ruins. In *The anti-aesthetic: Essays on postmodern culture*, ed. H. Foster. Seattle: Bay Press.

Cuomo, S. 2003. The sinews of war: Ancient catapults. *Science* 303:771–72.

Daniel, M. 2003. Struggling aquarium announces layoffs. Boston.com News, November 16.

Debord, G. 1977. *Society of the spectacle*. Detroit: Black and White.

DeLillo, D. 1986. *White noise*. New York: Penguin Books.

DeLoache, J. S., D. H. Uttal, and K. S. Rosengren. 2004. Scale errors offer evidence for a perception-action dissociation early in life. *Science* 304:1027–29.

de Saussure, F. 1983. *Course in general linguistics*. La Salle, IL: Open Court.

Donato, E. 1979. The museum's furnace: Notes toward a contextual reading of *Bouvard and Pécuchet*. In *Textual strategies: Perspectives in post-structuralist criticism*, ed. J. V. Harari. Ithaca, NY: Cornell University Press.

Eagleton, T. 1983. *Literary theory: An introduction*. Minneapolis: University of Minnesota Press.

Easthope, A., ed. 1993. *Contemporary film theory*. London: Longman.

Eco, U. 1976. *A theory of semiotics.* Bloomington: Indiana University Press.

———. 1979. *The role of the reader: Explorations in the semiotics of texts.* Bloomington: Indiana University Press.

———. 1984. *Semiotics and the philosophy of language.* Bloomington: Indiana University Press.

———. 1986. *Travels in hyperreality.* London: Picador.

———. 1995. *The island of the day before.* New York: Harcourt Brace.

———. 2000. *Kant and the platypus: Essays on language and cognition.* New York: Harcourt Brace.

Eisenstein, S. 1957a. *Film form: Essays in film theory.* New York: Meridian Books.

———. 1957b. *The film sense.* New York: Meridian Books.

Elgin, C. Z. 1983. *With reference to reference.* Indianapolis: Hackett Publishing.

———. 1997. *Between the absolute and the arbitrary.* Ithaca, NY: Cornell University Press.

Fisher, S. 2002. Objects are not enough. *Museums Journal* 102(6): 32–35.

Flaubert, G. 1976. *Bouvard and Pécuchet.* Harmondsworth, UK: Penguin Books.

Frankel, O. H., and M. E. Soulé. 1981. *Conservation and evolution.* London: Cambridge University Press.

Friedberg, A. 1993. *Window shopping: Cinema and the postmodern.* Berkeley: University of California Press.

Gass, W. H. 1978. *The world within the word.* New York: Alfred A. Knopf.

Gernsheim, H., and A. Gernsheim. 1956. *L. J. M. Daguerre (1787–1851): The world's first photographer.* Cleveland: World.

Ghiselin, M. T. 1997. *Metaphysics and the origin of species.* Albany: State University of New York Press.

Gonnerman, J. 2004. *Life on the outside: The prison odyssey of Elaine Bartlett.* New York: Farrar, Straus and Giroux.

Goodman, N. 1968. *Languages of art: An approach to a theory of symbols.* Indianapolis: Bobbs-Merrill.

———. 1976. Seven strictures on similarity. In *Problems and projects.* Indianapolis: Bobbs-Merrill.

———. and C. Z. Elgin. 1988. *Reconceptions in philosophy and other arts and sciences.* Indianapolis: Hackett Publishing.

Gorilla's amazing leap puzzles zoo experts. 2004. CNN.com, June 18.

Hagenbeck, L. 1956. *Animals are my life.* London: Bodley Head.

Hagoort, P., L. Hald, M. Bastiaansen, and K. M. Petersson. 2004. Integration of word meaning and world knowledge in language comprehension. *Science* 304:438–41.

Hannah, B. 1996. Carriba. In *High lonesome.* New York: Grove Press.

Hanson, E. 2002. *Animal attractions: Nature on display in American zoos.* Princeton, NJ: Princeton University Press.

Hanson, R. B., and F. E. Bloom. 1999. Fending off furtive strategists. *Science* 285:1847.

Harbison, R. 1977. *Eccentric spaces.* New York: Alfred A. Knopf.

Hardwick, M. J. 2004. *Mall maker: Victor Gruen, architect of an American dream.* Philadelphia: University of Pennsylvania Press.

Hargrove, E. 1995. The role of zoos in the twenty-first century. In *Ethics of the ark: Zoos, animal welfare, and wildlife conservation,* eds. B. G. Norton, M. Hutchins, E. Stevens, and T. L. Maple. Washington, DC: Smithsonian Institution Press.

Hassan, I. 1987. *The postmodern turn: Essays in postmodern theory and culture.* Columbus: Ohio State University Press.

Hasson, U., Y. Nir, I. Levy, G. Fuhrmann, and R. Malach. 2004. Intersubject synchronization of cortical activity during natural vision. *Science* 303:1634–40.

Hayles, N. K. 1995. Simulated nature and natural simulations: Rethinking the relation between the beholder and the world. In *Uncommon ground: Toward reinventing nature,* ed. W. Cronon. New York: W. W. Norton.

Hayward, S. C., and S. S. Franklin. 1974. Perceived openness-enclosure of architectural space. *Environment and Behavior* 6:37–52.

Hediger, H. 1969. *Man and animal in the zoo: Zoo biology.* New York: Delacorte Press.

Hershenson, M. 1999. *Visual space perception: A primer.* Cambridge, MA: MIT Press.

Hofstadter, D. R., and D. C. Dennett. 1981. *The mind's I: Fantasies and reflections on self and soul.* Toronto: Bantam Books.

Hull, D. L. 1989. *The metaphysics of evolution.* Albany: State University of New York Press.

Hyson, J. 2000. Jungles of Eden: The design of American zoos. In *Environmentalism in landscape architecture,* ed. M. Conan. Washington, DC: Dumbarton Oaks.

Iser, W. 1978. *The act of reading: A theory of aesthetic response.* Baltimore: Johns Hopkins University Press.

Jacobs, C. 2002. Reeling in the thrills. Sarasota Herald-Tribune, August 20, 1D-2D.

Jameson, F. 1983. Postmodernism and consumer society. In *The anti-aesthetic: Essays on postmodern culture,* ed. H. Foster. Seattle: Bay Press.

———. 1991. *Postmodernism: Or, the cultural logic of late capitalism.* Durham, NC: Duke University Press.

Jeanneret, C. E. ("Le Corbusier"). 1961. *Oeuvre complète* 1938–1946. Zürich: Les Editions Girsberger.

Joyce, J. 1961. *Ulysses.* New York: Vintage Books.

Kennedy, J. M. 2001. Smart geometry! In *Looking at looking: An introduction to the intelligence of vision,* ed. T. E. Parks. Thousand Oaks, CA: Sage Publications.

Kepes, G. 1965. Introduction. In *The Nature and art of motion,* ed. G. Kepes. New York: George Braziller.

Kisling, V. N. Jr. 2001. Ancient collections and menageries. In *Zoo and aquarium history; Ancient animal collections to zoological gardens,* ed. V. N. Kisling Jr. Boca Raton, FL: CRC Press.

Koh, L. P., R. R. Dunn, N. S. Sodhi, R. K. Colwell, H. C. Proctor, and V. S. Smith. 2004. Species coextinctions and the biodiversity crisis. *Science* 305:1632–34.

Kracauer, S. 1965. *Theory of film: The redemption of physical reality.* London: Oxford University Press.

Krasner, J. 1992. *The entangled eye: Visual perception and the representation of nature in post-Darwinian narrative.* New York: Oxford University Press.

Kuhn, T. S. 1970. *The structure of scientific revolutions,* 2nd edition. Chicago: International Encyclopedia of Unified Science 2(2).

Lakoff, G., and M. Johnson. 1980. *Metaphors we live by.* Chicago: University of Chicago Press.

Lapsley, R., and M. Westlake. 1988. *Film theory: An introduction.* Manchester, UK: Manchester University Press.

Lee, J. L. C., B. J. Everitt, and K. L. Thomas. 2004. Independent cellular processes for hippocampal memory consolidation and reconsolidation. *Science* 304:839–43.

Lee, K. 1999. *The natural and the artefactual: The implications of deep science and deep technology for environmental philosophy.* Lanham, MD: Lexington Books.

Linden, G. W. 1970. *Reflections on the screen.* Belmont, CA: Wadsworth.

Link, T. 1883. Zoological gardens, a critical essay. *American Naturalist* 17:1225–29.

Lipton, P. 2005. Testing hypotheses: Prediction and prejudice. *Science* 307:219–21.

Loomis, J. M. 2003. Visual space perception: Phenomenology and function. *Arquivos Brasileiros de Oftamologia* 66:26–29.

Lyotard, J-F. 1991. *The inhuman: reflections on time.* Stanford, CA: Stanford University Press.

Macbeth, N. 1971. *Darwin retried: An appeal to reason.* Boston: Gambit.

Major, C. 1975. Body heat. In *Reflex and bone structure.* New York: Fiction Collective.

Malamud, R. 1998. *Reading zoos: Representations of animals and captivity.* New York: New York University Press.

Marion, F. 1869. *L'Optique.* Paris: Librairie de L. Hachette et Cie.

Martel, Y. 2001. *Life of Pi.* New York: Harcourt.

McConnell, F. 1979. *Storytelling and mythmaking: Images from film and literature.* New York: Oxford University Press.

McKibben, B. 1989. *The end of nature.* New York: Random House.

McLuhan, M. 1969. *The Gutenberg galaxy: The making of typographic man.* New York: Signet.

McPhee, M. E. 2003. Effects of captivity on response to a novel environment in the Oldfield mouse (*Peromyscus polionotus subgriseus*). *International Journal of Comparative Psychology* 16:85–94.

Merchant, C. 1995. Reinventing Eden: Western culture as a recovery narrative. In *Uncommon ground: Toward reinventing nature,* ed. W. Cronon. New York: W. W. Norton.

Merrell, F. 1995. *Semiosis in the postmodern age.* West Lafayette, IN: Purdue University Press.

———. 1997. *Peirce, signs, and meaning.* Toronto: University of Toronto Press.

Metz, C. 1974. *Film language: A semiotics of the cinema.* New York: Oxford University Press.

———. 1982. *The imaginary signifier: Psychoanalysis and the cinema.* Bloomington: Indiana University Press.

Metzinger, T. 2003. *Being no one: The self-model theory of subjectivity.* Cambridge, MA: MIT Press.

Mishler, B. D. 1999. Getting rid of species? In *Species: New interdisciplinary essays,* ed. R. A. Wilson. Cambridge, MA: MIT Press.

Monaco, J. 1981. *How to read a film,* Revised edition. New York: Oxford University Press.

Muggeridge, M. 1974. *Chronicles of wasted time.* Chronicle 2, *The infernal grove.* New York: William Morrow.

Mullan, B., and G. Marvin. 1987. *Zoo culture.* London: Weidenfeld and Nicolson.

Nagel, T. 1974. What is it like to be a bat? *Philosophical Review* 83:435–50.

O'Brien, G. 1993. *The phantom empire.* New York: W. W. Norton.

Ogden, C. K., and I. A. Richards. 1960. *The meaning of meaning: A study of the influence of language upon thought and of the science of symbolism.* New York: Harcourt, Brace.

Pagels, E. 2003. *Beyond belief: The secret gospel of Thomas.* New York: Vintage Books.

Pamuk, O. 2004. *Snow.* New York: Alfred A. Knopf.

Peirce, C. S. 1966. *Charles S. Peirce: Selected writings (values in a universe of chance).* New York: Dover Publications.

Pessoa, L. 2004. Seeing the world in the same way. *Science* 303:1617–18.

Popper, K., and D. Miller. 1983. A proof of the impossibility of inductive probability. *Nature* 302:687–88.

Popper, K. R. 1965. *The logic of scientific discovery.* New York: Harper Torchbooks.

———. 1968. *Conjectures and refutations: The growth of scientific knowledge.* New York: Harper Torchbooks.

Riedl, R. 1984. *Biology of knowledge: The evolutionary basis of reason.* Chichester, UK: John Wiley and Sons.

Robertson, E. G. 1831. *Mémoires: Récréatifs, scientifiques et anecdotiques d'un physicien-aéronaute. I. La Fantasmagorie.* Paris: Café, Clima editeur.

Rothfels, N. 2002. Immersed with animals. In *Representing animals,* ed. N. Rothfels. Bloomington: Indiana University Press.

Salmon, W. C. 1995. Problem of induction. In *The Cambridge dictionary of philosophy,* ed. R. Audi. New York: Cambridge University Press.

Schama, S. 1995. *Landscape and memory.* New York: Alfred A. Knopf.

Schickel, R. 1968. *The Disney version.* New York: Simon and Schuster.

Sebeok, T. A. 2001. *Signs: An introduction to semiotics,* 2nd edition. Toronto: University of Toronto Press.

Seife, C. 2004. Gravity probe to give Einstein a pricey high-precision test. *Science* 304:385.

Serrell, B. 1998. *Paying attention: Visitors and museum exhibitions.* Washington, DC: American Association of Museums.

Shepard, P. 1995. Virtually hunting reality in the forests of simulacra. In *Reinventing nature? Responses to postmodern deconstruction,* eds. M. E. Soulé and G. Lease. Washington, DC: Island Press.

Sontag, S. 1961. *Against interpretation and other essays.* New York: Farrar, Straus and Giroux.

———. 1973. *On photography.* New York: Farrar, Straus and Giroux.

———. 2004. *Regarding the pain of others.* New York: Picador.

Sorkin, M. 1992. Introduction: Variations on a theme park. In *Variations on a theme park: The new American city and the end of public space,* ed. M. Sorkin. New York: Noonday Press.

Spotte, S. 1975. Jawsome. *Animal Kingdom* 78(4): 2–9.

———. 1992. *Captive seawater fishes: Science and technology.* New York: John Wiley and Sons.

———. 2002. *Candiru: Life and legend of the bloodsucking catfishes.* Berkeley, CA: Creative Arts Book Company.

Spotte, S., and P. Clark. 2004. A knowledge-based survey of adult aquarium visitors. *Human Dimensions of Wildlife* 9:143–51.

Spreiregen, P. D. 1965. *Urban design: The architecture of towns and cities.* New York: McGraw-Hill.

Sternberger, D. 1977. *Panorama of the nineteenth century.* New York: Urizen Books.

Stille, A. 2002. *The future of the past.* New York: Farrar, Straus and Giroux.

Teske, P. R., M. I. Cherry, and C. A. Matthee. 2003. Population genetics of the endangered Knysna seahorse, *Hippocampus capensis. Molecular Ecology* 12:1703–05.

Todorov, T. 1977. *The poetics of prose.* Ithaca, NY: Cornell University Press.

Ulmer, G. L. 1983. The object of post-criticism. In *The anti-aesthetic: Essays on postmodern culture,* ed. H. Foster. Seattle: Bay Press.

USA Today. 2004. Robo-dinosaurs coming to L. A. Zoo. May 24, 3A.

Vidler, A. 2000. *Warped space: Art, architecture, and anxiety in modern culture.* Cambridge, MA: MIT Press.

Wieseltier, L. 2004. Yehuda Amichai: Posthumous fragments. *New York Times Book Review* 109(47): 39.

Wilden, A. 1987. *The rules are no game.* London: Routledge and Kegan Paul.

Author Index

Page numbers in italics refer to illustrations.

Abbott, E. A., 142, 176 n. 1
Alfred, B. M., 178 n. 18
Amis, M., 91, 171 n. 1
Andrew, J. D., 100, 134, 171 n. 7, 172 n. 46, 176 n. 4
Arnheim, R., 124, 134, 150–51, 171 n. 27, 174 n. 1, 177 nn. 27–31

Balázs, B., 92, 171 n. 27
Bamford, C. L., 146, 177 n. 16
Barnouw, E., 112, 173 nn. 22 and 23, 25
Barthes, R., 98, 120, 124–27, 129–32, 139–40, 172 n. 29, 174 n. 3, 8, 12 and 13, 175 nn. 15–21, 36–40, 42–43, 176 nn. 22, 23, and 26
Batty, M., 177 nn. 18 and 19
Baudrillard, J., 17, 19, 37, 40–42, 54, 61–62, 74, 126, 164 nn. 1 and 2, nn. 16–19, 166 n. 2, 9 and 10, 15, 168 n. 40, 46, and 47, 169 n. 25, 174 n. 8
Bazin, A., 96–97, 171 n. 18, 20–22, 27, 172 n. 32
Benedikt, M. L., 14, 147, 164 n. 11, 177 nn. 17 and 20
Benjamin, W., 13–15, 21, 48, 51, 64–65, 90, 107, 121, 135, 142, 164 nn. 3, 6, 14, and 23, 167 n. 37, n. 6 (chap. 3), 168 nn. 48–49, 51, 52, and 53, 170 nn. 31 and 32, 171 n. 23, 174 n. 50, 176 n. 10
Bloom, F. E., 178 n. 18
Bordwell, D., 94–95, 102–3, 169 n. 3, 171 nn. 13, 15, 24, 25, and 26, 172 nn. 1–4, 173 nn. 10 and 12
Borgmann, A., 59, 168 n. 30
Bousé, D., 92–93, 171 nn. 6, 8, and 9
Brinkley, D., 176 n. 4

Burch, N., 101, 172 n. 48, 173 nn. 36 and 37

Calvino, I., 57, 168 n. 23
Chandler, D., 67–68, 70, 73–74, 164 n. 1 (chap. 1), 169 nn. 8, 13, and 21
Chandler, R., 67, 169 n. 4
Christopher, M., 173 nn. 22 and 24
Clark, P., 164 n. 22, 165 n. 20
Clubb, R., 177 n. 15
Cobley, P., 165 n. 30, 170 n. 5
Cohan, S., 66, 69–70, 169 n. 2, 9–12, and 14
Conway, W., 58, 60, 168 n. 25
Coover, R., *102*, 166 n. 17
Cortázar, J., 55–56, 166 n. 17, 167 nn. 16–19
Cox, D., 173 n. 9
Crawford, M., 50, 121–22, 167 n. 4, 173 n. 16, 174 nn. 51 and 52
Crimp, D., 134, 176 nn. 8 and 9
Cuomo, S., 166 n. 10

Daniel, M., 171 n. 2
de Saussure, F., 67, 98, 164 (chap. 1), 165 (chap. 1 nn. 1 and 3), 169 n. 7
Debord, G., 91, 129, 171 n. 3, 175 n. 30
DeLoache, J. S., 177 n. 24
Dennett, D. C., 38, 166 n. 4
DiLillo, D., *37*
Donato, E., 45–46, 165 n. 26, 166 nn. 24–27

Eagleton, T., 67, 73, 78, 154, 169 n. 6–7, 17, and 22, 170 n. 6 and 13, 177 n. 3
Easthope, A., 171–72 n. 27
Eco, U., 22, 25, 28–31, 40, 49, 52–55, 57, 71, 73–77, 80, 95, 121, 164

187

Eco, U. (*continued*)
  (chap. 1), 165 nn. 9–14, 17–19, and 22, 166 n. 8, 167 nn. 1, 7, and 12, 168 n. 22, 169 nn. 23 and 24, 170 nn. 1 and 3, 171 n. 19, 174 n. 49
Eisenstein, S., 171 n. 27
Elgin, C. Z., 72, 88, 126, 128, 162, 165 nn. 8 and 21, 166 n. 6, 169 n. 20, 170 n. 30, 175 nn. 23 and 24, 176 n. 19, 178 n. 31

Fisher, S., 120, 174 n. 46
Flaubert, G., 17, 88, 153, 158, 170 n. 29
Frankel, O. H., 162, 178 n. 28
Franklin, S. S., 146, 176 nn. 10 and 11, 177 n. 14
Friedberg, A., 14, 107, 116, 118, 164 nn. 4, 8 and 9, 173 nn. 14–21, and 34, 174 nn. 39, 40, 42, 44, 45, and 47

Gass, W. H., *49*
Gernsheim, A. and H., 113, 116, 173 nn. 28, 30, 31, and 32
Ghiselin, M. T., 165 n. 23, 166 n. 3, 170
Gonnerman, J., 176 n. 5
Goodman, N., 43, 62, 67, 72, 126, 128, 133, 166 nn. 21 and 22, 168 nn. 44 and 45, 169 nn. 5 and 19, 174 nn. 10 and 11, 175 nn. 23 and 24, nn. 2 and 3 (chap. 9)

Hagenbeck, L., 173 n. 35
Hagoort, P., 169 n. 16
Hannah, B., *91*, 166 n. 17
Hanson, E., 164 n. 21
Hanson, R. B., 178 n. 18
Harbison, R., 46–47, 166 nn. 29, 30, and 31, 167 n. 32
Hardwick, J. W., 64, 122, 168 n. 50, 174 nn. 53 and 54
Hargrove, E., 56–59, 168 nn. 20, 24, 27–29
Hassan, I., 169 n. 7, 170 n. 4
Hasson, U., 172 n. 8
Hayles, N. K., 48, 147, 167 n. 34, 168 n. 21, 177 n. 23
Hayward, S. C., 146, 176 nn. 10 and 11, 177 n. 14
Hediger, H., 173 n. 35

Hershenson, M., 177 n. 22
Hofstadter, D. R., 38, 166 n. 4
Hull, D. L., 39, 166 n. 7, 174 n. 6
Hyson, J., 60, 168 nn. 36 and 37

Iser, W., 72–73, 78–83, 169 n. 18, 170 nn. 7, 10–12, 14–22, and 24, 172 n. 38

Jacobs, C., 171 n. 4
Jameson, F., 14–15, 164 nn. 5, 12, and 13, 165 (chap. 1), 166 n. 16, 172 n. 37, 175 nn. 28 and 29
Jansz, L., 165 n. 30, 170 n. 5
Jeanneret, C. E. "Le Corbusier," 92
Johnson, M., 67–68
Joyce, J., 29, 77, 106, 165 n. 15

Kennedy, J. M., 177 n. 26
Kepes, G., 142, 176 n. 3
Kerouac, J., 134
Kisling, V. N. Jr., 164 n. 21
Koh, L., P., 178 n. 30
Kracauer, S., 171 n. 27
Krasner, J., 84, 170 n. 26
Kuhn, T. S., 169 n. 7

Lakoff, G., 67–68
Lapsley, R., 171–72 n. 27
Lee, J. L. C., 170 n. 8
Lee, K., 42, 47–48, 54, 166 nn. 19 and 20, 167 nn. 35 and 36, n. 14 (chap. 3)
Linden, G. W., 98–99, 103, 172 nn. 33, 35 and 36, nn. 5–6 (chap. 7), 173 n. 13
Link, T., 16, 164 n. 15
Lipton, P., 178 n. 19
Loomis, J. M., 164 n. 10, 177 n. 22
Lyotard, J-F., 164 n. 20

Macbeth, N., *66*
Major, C., *75*
Malamud, R., 167 n. 19, 174 n. 38
Marion, F., *111, 112, 115,* 173 nn. 26 and 27
Mason, G., 177 n. 15
Martel, Y., 62, 168 n. 43
Marvin, G., 52, 60, 167 nn. 8 and 9, 168 nn. 38 and 39, 171 nn. 16 and 17, 174 n. 38

Merchant, C., 167 n. 7
Merrell, F., 32, 36, 164–65 (chap. 1), 165 nn. 13, 24, 25, 28, and 29, 166 n. 18
Metz, C. 74, 94, 98–100, 140, 171 nn. 10, 11, and 14, 172 nn. 31, 34, 40–45, and 47, 175 nn. 15 and 22, 176 n. 27
Metzinger, T., 170 n. 25
McConnell, F., 90, 105, 170 n. 33, 173 n. 10
McKibben, B., 168 n. 33
McLuhan, M., 13, 83, 87, 166 n. 11, 170 n. 23
McPhee, M. E., 178 n. 32
Miller, D., 178 n. 23
Mishler, B. D., 162, 178 n. 29
Mitry, J., 100
Monaco, J., 98, 171–72 n. 27
Muggeridge, M., 137–38, 176 n. 17
Mullan, B., 52, 60, 167 nn. 8 and 9, 168 nn. 38 and 39, 171 nn. 16 and 17, 174 n. 38

Nagel, T., 61, 153, 168 n. 41

O'Brien, G., 94, 171 n. 12
Ogden, C. K., 23, 165 nn. 2 and 4

Pagels, E., 165 n. 27
Pamuk, O., 14, 164 n. 7
Peirce, C. S., 27, 32, 36, 77, 164 (chap. 1), 165 n. 3 and 8, 170 n. 5
Pessoa, L., 172 n. 8
Popper, K. R., 21, 154–55, 157–60, 177 nn. 1, 2, 4–6, 8 and 9, 178 nn. 10–15, 17, 19, 21, 23–27
Prévost, P., 113–14

Richards, I. A., 165 nn. 2 and 4
Riedl, R., 158, 178 n. 16

Robertson, É. G. R., 10–13, *113*, 173 nn. 22, 24
Rothfels, N., 60, 168 nn. 34 and 35

Salmon, W. C., 178 n. 20
Schickel, R., 54, 167 n. 13
Sebeok, T. A., 25, 165 nn. 5, 6, and 7
Seife, C., 177 n. 7
Serrell, B., 174 n. 46
Shama, S., 134, 176 n. 7
Shepard, P., 59, 168 nn. 31 and 32
Shires, L. M., 66, 69–70, 169 nn. 2, 9–12, 14
Smith, V. S., 178
Sodhi, N. S., 178
Sontag, S., 124–25, 127–29, 132–36, 138–39, 143, 167 n. 19, 174 nn. 4, 5, 7, 14, 175 nn. 25–27, 31–35, 44, 176 n. 6 (chap. 9), 176 nn. 6, 11–16, 18, 20, 21, 24, 25, n. 6 (chap. 10)
Sorkin, M., 50, 167 nn. 2 and 3
Soulé, M. E., 162, 178 n. 28
Spotte, S., 164 n. 22, 165 n. 20, 166 nn. 12, 13, and 14, 167 n. 11, 168 n. 39, 172 n. 39, 174 n. 4 (chap. 8)
Spreiregen, P. D., 145, 176 n. 10, 177 nn. 12 and 13
Sternberger, D., 116, 173 n. 33
Stille, A., 166 n. 5

Teske, P. R., 178 n. 22
Todorov, T., 66, 169 n. 1, 178 n. 26

Ulmer, G. L., 164–65 n. 1, 167 n. 15

Vidler, A., 46, 92, 103, 144, 147, 166 n. 2, 171 n. 5, 172 n. 7, 176 nn. 7, 8, and 9, 177 n. 21

Westlake, M., 171–72 n. 27
Wieseltier, L., 168 n. 26
Wilden, A., 84, 165 n. 29, 170 n. 27

Subject Index

Page numbers in italics refer to illustrations.

8½, 107
abduction, 28
Adams, A., 59
Adams, J., 113
Adventures of Lucky Pierre, The: Directors' Cut, 102
Africa, 58
African lion, 93
African Queen, 54
African savannah, 52
African veldt, 63–64
Aladin [*sic*] *and His Wonderful Lamp,* 114
Alaska, 38
Alpine ibex, 26, 28, 31–33, 35–36, 38, 45, 60, 72, 78, 94. See also *Capra ibex*
American bison, 47
American slang, 170 n. 9
Americans, 176 n. 5
Amichai, Y., 58
"animal rightists," 174 n. 41
animals: alternative existence and, 74; as actors, 95; as artifacts, 15, 45, 162; as commodities, 121; as curiosities, 17, 21, 107, 123; as ecosystem fragments, 143; as fragments, 45–48; as individuals, 37, 45; as photographed objects, 127, 141; as spectacle, 44; as visual objects, 151; behavior of, 16, 99; cinematic images and, 55; endangered, 15, 17, 19–20, 26, 48–49, 143; genetic manipulation of, 15, 48; Hagenbeck's legacy to, 168 n. 34; hiding of, 97; humans and, 104, 167 n. 8; "immersion" exhibit and, 61; in real time, 91; indelibility of, 29; information and, 16; memory and, 152; metaphorical subsumation of, 84; motion of, 118; myth and, 74; naming of, 44–45, 52, 167 n. 10; objectification of, 167 n. 19; personification of, 52, 153; in postmodernism, 22, 139, 153; reality and, 93; self-image and, 135; semiotically real, 44–45, 84, 123, 136, 138, 162; spatial context of, 143–46; spectator and, 120–21; taunting of, 95; theme parks and, 50; vs. simulated Nature, 64; wildlife films and, 94; zoo graphics and, 73; zoo space and, 99
Antarctic icepack, 144
antonym, 69
aquariums: history of, 164 n. 21; modernism and, 13, 15; motion in, 118
arcade: adaptation by zoos, 123; evolution of, 108; *flâneur*'s gaze and, 107; Gruen and, 122; modernist, 109; Parisian, 107
Arcadian gardens, 49
architecture: film and, 92; isovist in, 147; landscape, 60; painting of, 114; postmodern, 147; shopping mall and, 122; theatrical contrast and, 96; zoo, 144–45
Arctic Circle, 144
Arctic tern, 144
Aria de Capa, 58
Arial, 87
Aristotle, 94–95
Aristotle's *Poetics,* 94
art: abstract, 138; artifacts and, 54; communication by, 98; conflict in, 82; criteria for, 134; definition of, 124; frame and, 57; in zoos, 21; modernism and, 13; motion and, 92; perceived reality of, 44; perception of

art (*continued*)
  reality and, 100; perception of, 134; photographic seeing and, 139; photography as, 124, 133–35; pop, 136; reality and, 129; representation of reality and, 62; semiotic value and, 40; storytelling and, 128; Surrealism movement and, 136; transcendence in, 99; transformation of objects and, 134; zoo exhibit as, 124
Art and Visual Perception, 134
art theory, 20
artifact: animals as, 162; definition of, 42; diorama as, 64; in art, 54; in zoo exhibit, 60, 136; natural kinds and, 48; photographs as, 135; reality and, 54; science and, 45; "second-order" simulations as, 65; signification by, 46; transcendence by, 64; zoo animals as, 45, 47–48
artificial insemination, 20
As I Lay Dying, 107
Asia, 58, 160
Asian economy, 160
Athens, 166 n. 6
Audubon magazine, 138
Australia, 159
Australian aborigine, 70

Bahia Mar, 76
Bambi, 74
Bambi, 106
Barker, R., 113
Barth, J., 166 n. 17
Barthes' *noeme,* 127–28
Barthes' *punctum,* 130–31
Barthes' *spectrum,* 127
Barthes' *studium,* 130–31
Barthes' *unary,* 139
Bartlett, E., 176 n. 5
Baudelaire, C., 13, 96, 107
Bentham, J., 118, *119,* 174 n. 41
Bering Sea, 167 n. 10
Berlin Industrial Exhibition, 117
biodiversity, 161–62
biological species definition, 160–61
bird shows, 153
black-footed ferret, 47
Boodles gin, 76

Bordwell's *schemata,* 102, 120, 172 n. 2
Borges, J. L., 37, 166 nn. 1 and 17
Boston, 113, 171 n. 2
Boston Red Sox, 37–38
Bourbon Street, 51
Bouton, C. M., 114
Bouvard, 88, 153, 158
Bouvard and Pécuchet, 153
Bovidae, 28, 31
brain: other humans and, 103–4; processing statements by, 169 n. 16; synchrony in film watching and, 104; visual perception and, 147
Bruce (the white shark), 55
Busch Gardens, 171 n. 4
Busted Flush, 76

California, 53
California condor, 48
camera: beautifying by, 138; cryptic animals and, 97; in film theory, 103; in myth making, 138; in zoo photography, 137; photographic seeing and, 139; reality and, 129, 139; sublimation to gun and, 129; transformation of history and, 127
camera angles, 95, 97, 101, 117
camera lens: as death's witness, 125; as time's substitute, 126; in early film theory, 95, 97
Camera Lucida: Reflections on Photography, 124
Canadian Rockies, 120
Cancún, 42
Candiru, 41
Canis familiaris, 29–30
Canterbury Cathedral, 114–15
Capra ibex, 26, 31, 38, 94. *See also* Alpine ibex
Capuchin monastery, 110
Casey (at the bat), 94
causality: in film theory, 105; in stories, 66; in zoo text, 86. *See also* plot
Charles W. Morgan, 39
Che Guevara, 175 n. 22
China's Dafeng Reserve, 167 n. 33
Chronicles of Wasted Time, 137
cinema: as commodity experience, 121; Barthes' *noeme* and, 127; definition

of, 174 n. 2; development of, 120; diorama as protocinema and, 116; fictive time and, 99; forgery of, 134; mobility and, 123; motion and, 91, 100; "Nature" and, 49; realism and, 17; shop window as, 110; shopping mall and, 109; spectator participation and, 104; static exhibit and, 91; unseen seer and, 119; virtual gaze and, 107; virtual reality and, 14; vs. theater, 98–99. *See also* film; movies
cinema screen: magic lantern and, 112; shop window as, 110; viewing distance of, 117
cinematic time, 54
class: denotation and, 35, 138; interpretant and, 31; representational thinking and, 58; scientific theory and, 160; vs. individuals, 33, 37–38
Cleveland, Ohio, *109*
closed text: definition of, 76; in cinema, 92; in cinematic fiction, 106; model reader and, 77; zoo graphics as, 89. *See also* model text; open text; text
closed-text cinematic format, 92, 98, 106
cognition, 102–4
cognitive acquisition, 121
cognitive type (CT), 29–30, 71–72
collateral sulcus, 104
Columbian Museum, 113
Columbus, C., 50
commodification: cinematic image and, 135; in postmodernism, 15; connoisseurship and, 121; shopping mall and, 122; zoos and, 64
commodities: animals as, 121; enhancement by animals and, 122; in postmodernism, 135; Marx on, 120; mobilized gaze and, 108; services as, 121, 135; tourism as, 119
complex knowledge, 30–31
computer games, 19
connotation: definition of, 34; enrichment by, 84; examples of, 70; in fiction, 78; in real time; in wildlife films, 93; objects and, 34; replacement by denotation and, 83; symbols in, 34
conservation: as metaphor, 45; as propaganda, 82, 125; "immersion" exhibit as, 60; message in, 61; vs. preservation, 161–62; zoo graphics and, 88; zoos and, 17, 20–21, 43, 121, 161–62
constructivist psychology, 102–3
counterfeit simulation. *See* "first-order" simulation
Creationists, 158
Crocus, 134
culture: signs and, 17; stories and, 66, 93; wildness and, 52
Cupid, 135
cyberspace; ageographical city in, 167 n. 2; information in, 19; virtual reality and, 50
Cygnus atratus, 159

Daguerre, L. J. M., 61, 113–14
daguerreotype, 133
Dallas Zoo, 61
Dalmatian dog, 147, *148*
Dammtor, 117
Dan'l Webster (the frog), 70
Darwin Retried: An Appeal to Reason, 66
David, J-L., 173 n. 29
de Philipstahl, P., 113
Decimus Burton's Colosseum [*sic*], 116
découpage, 92
deduction, 159
Deep Blue Good-by, The, 76
"deep ecology," 165 n. 27
Degas, E., 151, *152*
Democratic National Convention (1968), 42
Dendrocygna javanica, 86
denotation: by words, 79; definition of, 27, 35; description as, 78; diegesis and, 44, 171 n. 13; imaginary animals and, 126; "immersion" exhibit and, 62; in text, 84, 86, 89; in zoo text, 83–84, 88; individual and, 138; narrative and, 35–36, 86; Nature and, 93; of reality, 100; representation and, 62, 126; signification and, 84; synonyms and, 86; wandering viewpoint and, 83

Denver, *96, 110*
Denver Public Library, *96, 110*
department store, 108
description: as science, 178 n. 26; by language, 128; denotation and, 78; of zoo graphics and text, 67, 80–82, 84–85, 88; photograph and, 128; while touring, 120
determinant judgment, 28. *See also* judgment; reflective judgment
diegesis: definition of, 44, 171 n. 13; history of, 171 n. 13; in film, 44, 99–100; in protocinematic entertainments, 120; narrative and, 94
diorama: as artifact, 64; as protocinematic entertainment, 110, 114–16, 120; development of, 19, 116; etymology of, 116; "immersion" exhibit and, 61; in zoos, 56; painting and, 148; virtual gaze and, 107; working of, *115*
discourse: flow of, 71; function of, 66; narrative as, 73
Disney: as hyperreal, 42, 50, 60; narrative experience and, 59; sign and, 26
Disneyland: as hyperreal, 49, 53: film studio and, 54
dissimulation, 61. *See also* simulation
Dolly (the sheep), 19
dolphin shows, 15, 122, 153, 171 n. 4
doxa, 33, 162
dramatic place, 95. See also *locus dramaticus*
Duchamp, M., 92
Dumbo (the elephant), 58

E. T. (the alien), 58
Earth, 38, 59, 142, 144
Eastwood, C., *104,* 172 n. 8
Eciton burchelli, 178 n. 30
Eden, 17, 21, 49, 62
Edina, Minnesota, 122
Edinburgh, 113
education: *Fantasmagorie* and, 113; in zoos, 20, 21; "naturalistic" exhibit and, 43; sign and, 33; zoo text and, 80
Edward and Victoria (film characters), 106
eidos, 22

Einstein's theory of general relativity, 177 n. 7
elevator, 116
Elizabethan English, 170 n. 9
ellipsis, 92
Elsa, 52
empiricism: fiction and, 78; in science, 154, 156–57, 159; in sentences, 72; symbols and, 78
England, 113
English, 30, 71, 113
English Fleet Anchored . . . , The, 113
Enoshima Island, Japan, *53*
epistēmē. See knowledge
esthetic object. *See* textual meaning
essentialism, 165 n. 6
Europe, 167 n. 33, 174 n. 43
European Alps, 26, 32, 34, 72
event: description as, 128; false prediction of, 99; in cognitive psychology, 102; in fiction, 78–79; in narrative, 72, 74, 99; in novels, 73; in open text, 77; in scientific hypothesis testing, 157; in stories, 106; photograph and, 125; press and, 14; semiotic, 32; visual experience of, 148
evolution: balance (gravitational) in, 150; conservation and, 162; in zoos, 17, 162; natural kinds and, 48; of humankind, 166 n. 10
evolutionary fitness, 14
evolutionary theory, 46
exhibition hall, 107
existential statements, 159–60
experience: actualization of text and, 82; artistic, 134; at zoos, 120; atrophy of, 13; cognitive type and, 71; commodity and, 120–21; complex knowledge and, 31; connotative text and, 84; conscious, 170 n. 25; esthetic, 124; film theory and, 102–3; humans vs. animals and, 167 n. 8; hyperrealistic, 54; inductive statements and, 159; knowledge and, 156, 158; model reader and, 87; molar content and, 30; narrative and, 74, 80; non-affective, 98; novel, 99; of diorama, 116; of Nature, 168 n. 21; of reader, 67; overdetermined narrative and,

82; perspective in fiction, 81; phenomenal, 84; photography and, 127; recognition and, 27–30, 35; spectacle and, 44; to "read" zoo exhibits, 80; visual perception and, 147–48
expository writing, 73–74
eye: dispersal in film theory, 103; in art, 134; in photography, 136; perception by, 143; perspective and, 139, 142; photon sensing by, 176 n. 2
eye-line match, 92

"fablephobia," 88
fables, 89
fairy tales, 89
Fantasmagorie, 110, 112–13
Faulkner, W., 107
Faulknerian prose, 90
Fellini, F., 107
fiction: cinematic, 92, 94, 100, 106; denotation in, 93; familiarity in, 74; literary, 106; meaning in, 80; narrative in, 13, 72–73; participant and, 44; passivity and, 87; perspective in, 81; semiotics of, 78; zoo graphics and, 89
Fifth Avenue, 122
film: 180-degree line, 97; actualization of, 98; as closed text, 76; as imitation, 94; as language, 98, 172 n. 30; brain and, 103–4; cognition and, 102; definition of, 174 n. 2; diegesis in, 44, 99–100; documentary as narrative and, 92; esthetic of, 107; history and, 14; in postmodern culture, 92, 107; *locus dramaticus* and, 97; modernism and, 13; narrative event and, 99; narrative structure of, 107; painting and, 148; passivity and, 87; silent, 95, 120; spatial constraints and, 100–101; spectator participation and, 97–98; stories and, 93; theory of, 20, 94–95, 97–98, 102–5, 171 n. 13; "thirty degree rule" and, 117; vs. literature, 98, 105–6; wildlife and, 92–94, 97; zoos and, 21. *See also* cinema; movies
Film Technique, 95
Finnegans Wake, 77, 106–7, 170 n. 4

"first-order" simulation: ancient humans and, 166 n. 10; definition of, 40–41; examples of, 42; hyperreality and, 61; in theme parks, 54; zoos and, 49
flânerie, 107, 109–10, 120
flâneur, 107
flâneuse, 108
Flatland, 142–43
Flatland, 142–43
flicker fusion, 103
Flipper, 52
Florida, 76
Florida's fishing laws, 69
Fort Lauderdale, Florida, 76
Foucault, M., 174 n. 38
fragments: in art, 54; in postmodernism, 164–65 n. 1; of ecosystems, 144; representation and, 45–46
Fram, 117
France, 113, 134
Froggy, 70
fusiform gyrus, 104

gambling casinos, 15
García-Márquez, G., 166 n. 17
gaze: definition of, 107
goblins, 43
God, 62, 165 n. 27
Goethe, J. W. von, 65
Good, the Bad, and the Ugly, The, 104, 172 n. 8
Goodall, J., 52
grammar, 68, 128
Greece, 95
Greek mythology, 77
Gruen effect, 122
Gruen, V., 64, 122

H/D ratio, 145, *146*, 176 n. 11
Hagenbeck, C., 117, 145, 168 n. 34
Hagenbeck's Animal Park, 168 n. 34
Hamburg, 117
Hancock, J., 113
Hannah, B., 166 n. 17
Hans (the Alpine ibex), 26, 31, 33–36, 38, 94
Harvard Yard, 162–63
Hawai'i, 38

Haydn's symphony, 133
Heiligengeist Field, 117
Heraclitus, 161
Hercules, 73
High and Low, 101
Hindi, 86
history: at shopping mall, 50; film and, 14; fragments and, 46; in photographs, 127; of diegesis, 171 n. 13; of mimesis, 171 n. 13; of zoos and aquariums, 164 n. 21; postmodernism and, 13–14; television and, 14
Hollywood, 100
holography, 42
Homo sapiens, 31, 94
homology, 172 n. 30
homonym, 69
Hoolihan, M., 91
Hornor, T., 116
How to Read a Film, 98
human beings (humankind), 129, 153, 165 n. 27, 166 n. 10, 167 n. 8
human faces, 104, 128
Hungary, 140
hyperreality: definition of, 17; Disney as, 60; Disneyland as, 49, 53; "first-order" simulation and, 61; in postmodernism, 19; modernism and, 50; shopping and, 122; space in, 145; theme parks and, 54; "third-order" simulation as, 40–42; vs. realism, 59; zoos and, 49, 52
hypothesis: in film theory, 102–3, 105; tests of, 23. *See also* scientific hypothesis

icon: as nuclear content, 30; as sign, 26; as zoo logo, 127; definition of, 25; of American southwest, *18;* Peircean, 165 n. 8; Peircean Secondness and, 77; photographic image as, 33; proposition (compound symbol) and, 36; representational thinking and, 56; resemblance and, 43
ideation. *See* signification
Illuminations, 142
illusion: hyperreality and, 49, 53; "immersion" exhibit and, 63; in art, 99; in fiction, 79; literature as, 72, 169 n. 17; mental images as, 147; of consensus, 143; of novelty, 64; of reality, 100; of speed, 93; painting and, 151; panorama and, 113; photograph and, 100; press and, 14; shop window and, 122; spatial, 145; theater and, 95; virtual reality as, 50
image: 3D, 142; actualization of text and, 82, 84; as idea, 133; as moiety, 134; as praxis, 120; as sign, 36; as simulacrum, 37; bodily, 103; by fiction, 79; cinematic, 55, 87, 92, 125, 135; "commodity reification" and, 91; film narrative and, 99–100; hyperreality and, 41; incomplete objects, 147–48, 152; in constructivist psychology, 103; in postmodernism, 13–14, 129, 135–36, 147; indirect commodification and, 122; in zoo guidebooks, 43; machine-produced, 135; magic lantern and, 112; mental, 147; ontology of, 92; perceived in art, 124; perception of, 176 n. 2; photographic, 14, 124–25, 127–31, 133, 135–38, 162; power of, 17, 62; qualities of, 37; reality and, 15, 91; recognition of, 28; representational thinking and, 56; reproduction of, 65; schizophrenia and, 129; sign and, 128; spectator and film, 97; speech and, 172 n. 31; still, 91; streaming, 14, 98; surrealistic, 136; verbal description of, 128; visual perception of, 147–48
imagistic scene, 84
Imax theater, 91, 171 n. 2
imitation, 94
"immersion" exhibit, 60, 62–65, 144
index: as nuclear content, 30; as sign, 26, definition of, 25, proposition and, 36; Peircean, 165 n. 8; Peircean Secondness and, 77
indirect commodification, 15, 122
individual: as artifact, 45; balance (gravitational) and, 150; definition of, 37, 86; denotation of, 94, 138; metaphysics and, 166 n. 3; naming of, 44–45; Nature and, 45; preservation and, 162; recognition of, 28–29,

31, 35; representational thinking and, 58; scientific theory as, 160; species as, 165 n. 23; vs. class, 37–39
individualism, 15, 39
induction, 158–60
inductionism, 99, 158
inductive inference, 159
inductive statements, 159
industrial revolution, 40
inferior prefrontal cortex, 169 n. 16
infinite regress, 21, 154, 158
information: as *doxa,* 33; complex knowledge and, 30; control by description and, 81; in cyberspace, 19; in narrative, 90; in language, 71; model text and, 75; in postmodernism, 13; in scientific hypothesis testing, 157; spatial, 145; spectator and, 87–88; "third-order" simulation and, 40; transmission of, 90, 124; zoo graphics and, 72; zoos and, 16
insane asylum, 118, 174 n. 38
interactive exhibit, 91
Interior of Trinity Chapel Canterbury Cathedral, The, 114
international exposition, 121
interpretant: connotation and, 84; context of, 27; definition of, 24; denotation and, 83; in cognitive psychology, 102; in fiction, 79, 81; in film, 98; in graphics, 31; in language, 71; in semiotic progressions, 78; in spectacles, 94; in zoo text, 86; knowledge and, 27–28; meaning and, 35; Peircean, 69, 77; recognition of, 34; schizophrenia and, 129; semantic paradigms and, 70; signification and, *27,* 34; sign and, 25, 32; taxonomy and, 31; textual device and, 67. *See also* signified
invisible observer, 95
Irish, 113
Ise Shrine, 39, *40*
Island of the Day Before, The, 22
isovist, 147
isovistic field, 147, 177 n. 19
Iwo Jima, 175 n. 22

Jabari, 61
Japan, 39

Japanese, 70
Jacqueline Rocque, 105
Jardin des Plantes, 56
Jaws, 55, 101, 121
Jefferson, T., 113
Jeanneret, C. E. ("Le Corbusier"), *57*
Judgment, 28. *See also* determinant judgment; reflective judgment
jungle, 61, 86
Jurassic Park, 42, 55

Kafka, F., 86
Kant, E., 28, 31
Kant's *schemata,* 28–29, 172 n. 2
Katase Nishihama coast, *53*
Kennedy, J. F., 42
Kermit (the frog), 70
Kertész, A., *140*
Khedoori, T., 144
King, M. L. Jr., 42
Kinney's Shoes, 50
Kirk, Captain J., 38–39
Klein, W., 126
knowledge: anticipated, 75; brain and, 169 n. 17; cognitive type and, 71; in film theory, 102–3; in visual perception, 147; infinite regress and, 21, 154; information and, 16–17; Kant's *schemata* and, 28; model reader and, 87, 90; photograph and, 134; "reading" zoo exhibits and, 80; representation and, 31; semantics and, 169 n. 16; sign and, 23, 34; theoretical, 156, 158; zoo graphics and, 27, 72
Kurosawa, A., 101, 117–18

La Vaux's menagerie, 174 n. 38
La Vie au Pôle Nord, 117
Lacan, J., 129
Lady in the Lake, The, 67
Laputans, 67
language: as description, 128; as symbol system, 25; image and, 100, 128, 172 n. 31; in graphics, 72; Joyce and, 77; linguistic code and, 170 n. 2; literature and, 73, 78; meaning and, 71; means of imitation and, 94; metaphorical nature of, 67; metaphors and, 58; of fiction, 78; of film, 98,

language (*continued*)
  172 n. 30; schizophrenia and, 129; semiotics and, 23; signifier and, 67, 69; sign and, 71; translation of, 90
Las Vegas, 42
Lavoisier's *Chemistry,* 169 n. 7
Lawrence of Arabia, 172 n. 30
Le Figaro Illustré, 108
learning, 84
Leather World, 50
legend, 33, 74
Liège, 110
Life of Pi, 62
linguistic code, 75, 170 n. 2
linguistics, 71, 171 n. 14
Linnaean hierarchy, 162
literary theory: heroic selfhood and, 131; semiotic progression and, 78; spectacle and, 66; spectacle in, 169 n. 3, 171 n. 13; zoos and, 20. *See also* narrative theory
literature: artifacts and, 54; as illusion, 72, 169 n. 17; as mental instantiation, 134; conflict and, 82; connotation in, 84; didactic, 82; film and, 92; frame and, 57; Joyce and, 77; meaning in, 78; participation and, 104; performative nature of, 78; propagandist, 82; rhetorical, 82; speech and, 73; vs. film, 105–6
Little Red Book, 83
locus dramaticus, 95. *See also* dramatic place
logic, 154–55, 157–58
logical statement, 155–56
London, 37, 46, 113, 115–16, 174 n. 43
Lonely Silver Rain, The, 76
Los Angeles, 42, 106
Los Angeles Zoo, 171 n. 2

MacDonald, J. D., 76–77, 106
magic lantern, 110, 112, 120
magic show, 110
Magritte, R., 22, *23,* 36, *63,* 137
Man's Fate, 63, 137
Mao, Z., 83
Marinetti, F. T., 92
Marion, F., 112
Mars, 38–39

Marx, K., 120, 135
Mayan civilization, *18*
Mayday (1959), 126
McGee, T., 76, 106
meaning: brain and, 169 n. 16; denotation of, 84, 86; description and, 91; expository writing and, 73; in fiction, 78; in scientific theory, 158; in stories, 66–67; interpretant and, 24; narrative and, 74; nuclear content and, 30; in open text, 77; overdetermined narrative and, 82; paradigmatic structure of, 70; postmodernism and, 164–65 n. 1; reality and, 82; semantics and, 70–71; sign and, 34–35, 87; signification and, 25; signifier and, 67–68; syntagmatic structure and, 71; visualization by reader, 83
means, 94
Memento, 107
memory: biochemical basis of, 170 n. 8; fiction and, 78, 83; film and, 78; film images and, 100; images and, 17; individualism and, 39; language and, 71; Peircean Firstness and, 77, 79; Peircean Thirdness and, 79; postmodernism and, 13; reality and, 99; referent and, 140; shapes and forms and, 152; sign and, 17, 35
menagerie, 13, 163
metaphor: as icon, 25; examples of, *68,* falseness of, 88; fragments as, 45; Goodman's definition of, 67; idea as, 67; in film, 98; in self negation, 172 n. 36; language and, 58, 67; literal interpretation of, 72; of "ark," 143, 168 n. 34; painting and, 150; textual device and, 67; zoo text and, 80, 86
metonymy: as textual device, 67; in art, 54; in film, 98; in zoo text, 86; of fragments, 45; photography and, 139–40
Mexican dungeon, 113
Middle Ages, 56
Millay, Edna St. Vincent., 58
mimesis, 94–95, 171 n. 13
Mind's I, The, 38
mise-en-scène, 92, 110

moated exhibit, 117, 121, 137, 145, 173 n. 35
mobility: cinema and, 123; home entertainment and, 109; of *flânerie*, 110; of postmodern spectator, 120, 174 n. 41; protocinematic entertainments and, 108
mobilized gaze, 108–10, 120, 174 n. 41
mode, 94
model reader, 75, 77, 87, 90, 98
model text, 75, *See also* closed text; open text
modern age. *See* modernism
modernism: duration of, 164 n. 20; end of, 19; "high art" in, 54; hyperreality and, 50; landscape painting in, 56; mobility in, 120; "naturalistic" exhibit and, 60; objects and, 47; proselytism and, 162; shopping and, 173 n. 16; space and, 144; spatial perception and, 142–43; *trompe l'oeil*, 134; vs. postmodernism, 164 n. 20, 164–65 n. 1; zoo graphics and, 66, 90–91; zoo spaces and, 147; zoos, 13, 19, 21, 37, 59, 107, 129, 153, 163
molar content (MC), 30
Mona Lisa, 40, 45
montage, 92–93, 100, 123
Morris, R., 113
Moscow, 126
motion (movement): apparent, 103; cinema and, 91–92, 100; cinematic, 117; illusion in painting and, 151; Newton and, 150; object vs. copy and, 141; time and, 91; zoo exhibit and, 44, 91, 93, 118, 123, 136
Mouse, M. 26, 58, 60
movies: 4D, 171 n. 4; definition of, 174 n. 2; narrative in, 100; photography and, 128; theme parks and, 50, 55; time and, 15; wildlife photography and, 126. *See also* cinema; film
Mrs. Dalloway, 92
museums: art and, 134; artifacts and, 45–47; knowledge and, 134; protocinematic entertainment and, 107; shopping mall and, 64; wax, 49, 52, 55; window displays and, 120
music, 77, 133–34

Mystic, Connecticut, 39
Mystic Seaport Museum, 39
myth, 33, 74, 138

naïve realism, 14, 147
Nansen, F., 117
Narration in the Fiction Film, 94
narrative: as syntagmatic structure, 70; *Bambi* as, 106; characterization of, 74; cinematic, 92, 100; communication and, 58; construction of, 90, 105; definition of, 66; denotation in, 35–36; diegesis in, 44, 94; elements of, 74, 81; event and, 72, 74, 99; expository writing and, 73; in film theory, 97–98, 103; flow of, 82; fragments and, 144; knowledge and, 138; meaning from, 82; mimesis in, 94; motion and, 91; overdetermined, 82, 86; photograph as augmentative and, 175 n. 22; protocinematic entertainments and, 120; sign and, 66–67; stories in, 66–67; storytelling and, 72; structure of, 107; textual devices in, 67; touring and, 120; vs. photography, 128; zoo exhibit and, 44, 59, 65; zoo graphics and, 66, 86; zoo text and, 80, 86, 88, 106, 136
narrative eye, 84, 106
narrative theory, 84, 169 n. 3. *See also* literary theory
narrative thinking, 59–60
NASA, 177 n. 7
National Geographic magazine, 138
natural environment, 21
natural kinds, 48
"naturalistic" exhibit: as stage set, 171 n. 16; "immersion" exhibit and, 60; simulation and, 43; zoos and, 20, 44, 55–56
natural selection, 162
Nature: theater and, 49; as simulacrum, 63–64; as spectacle, 56; assault on, 143; beauty in, 13, 65; definition of, 19; denotation of, 93; Disneyland and, 54; distance and, 165 n. 26; film and, 92; fragments and, 46; future zoos and, 163; "hidden," 136; hyperreality and, 41; "im-

Nature (*continued*)
mersion" exhibit as, 60–62, 65; naming and, 45; "naturalistic" exhibit as, 55; objectification of, 168 n. 21; photography and, 124, 138; questions of, 90; rendered unthreatening, 52; representation and, 43, 49; representational thinking and, 58; science and, 158; shopping mall and, 50–51; signification and, 62; sign and, 65; simulated, 49, 53, 97, 142; space and, 144; spectator's gaze and, 44; Surrealism and, 137; theater and, 49; understanding of, 28, 84; vs. zoo exhibit, 135–36; zoo animals and, 162; zoos and, 20, 42–43, 153

negative allusion, 86

New England Aquarium, 171 n. 2

New Orleans, 51

New Times Roman, 87–88

New York *Post*, 113

New York, 42, 97, 113, 122

New Yorker magazine, 20

news media, 42

Newton, I., 150, 161

Night Train, 91

nightclubs, 15

Nolan, C., 107

nonfiction, 72

North America, 113

northern fur seal, *131,* 167 n. 10, 175 n. 41

novel: actualization of, 79; event and, 73, 99; fictional perspective of, 81; iterative, 76; MacDonald's 76; memory and, 78; storyteller and reader and, 44; vs. film, 98, 107

novelty, 15, 64, 81

nuclear content (NC), 30

Nude Descending a Staircase, 92

objects: artifactual, 43, 64; as art, 54, 124; as classes, 38; as individuals, 37; as metaphor, 67; as real, 128; as signs, 45; as simulacrums, 127; as spectacle, 44; attributes of, 34; aura of, 142; consensus and, 143; contingency of, 126; denotation and, 27; detail of, *149;* dynamic, 36; fragments of, 45–47; grammatical, 67; in modernist theory, 144; in photography, 133, 136; memory of, 152; motion and, 141; Nature as, 168 n. 21; or referent, 54, 78; painting and, 150–51; perceived reality and, 43, 64; perception of, 143, 145, 150; personified, 129–30; photographic transcendence of, 127, 139; photographs as, 136; qualities of, 36–37; recognition of, 28, 31; representation and, 45, 62; representational thinking and, 56; reproduction of, 65; science and, 154; semiotically real, 33; signification by, 45–46; simulation of, 37, 91; transformation into art, 134; visual density of, 148, *149. See also* semiotically real objects

observation, 154–59, 161

Oehler, A., 113

Old Arcade, The, *109*

On Photography, 124

open text: definition of, 77; plot in, 106; subjectivism in, 80; zoo graphics as, 89. *See also* closed text; model text; text

Opéra, 114

organism, 37, 39, 48

Orion, 37

Orlando, Florida, *44*

Orwell, G., 90

Other, the. *See:* Nature

O'Toole, P., 172 n. 30

painting: as mental instantiation, 134; autographic, 133; balance (artistic) in, 150–51; daguerreotype and, 133; dimensional space in, 134; diorama and, 114–16; film and, 92; iconographical tradition and, 175 n. 1; imitation and, 94; in theater, 95; landscape, 56, 59, 64; Magritte, 22, *23, 36, 63,* 137; of zoo panoramas, 117; panorama, 113; perception of, 134; Picasso, 54, *55, 105;* postmodernism and, 150; representation and, 43; Rousseau, 49; static image and, 95; storytelling and, 128; vs. photography, 125, 133–35, 175 n. 1

panopticon, 118, *119*, 174 n. 41
panorama: as protocinematic entertainment, 110, 113–14, 117, 120; development of, 19, 120; hydraulic lift and, 116; painting and, 148; photographic seeing and, 128; virtual gaze and, 107
paradigm: cinematic, 172 n. 31; definition of, 67–69, 169 n. 7; in scientific theory, 169 n. 7; meaning in linguistics, 169 n. 7; phonetic, 70–71; semantic, 70; syntactic, 70–71; television and, 91; zoo graphics and, 72; zoo text and, 86
Paris, 56, 97, 107, 110, 113–14, 117, 172 n. 41
Parole in litertà, 92
Peasant Girls Bathing in the Sea at Dusk, 151, *152*
Pécuchet, 88, 153, 158
Peircean Firstness: definition of, 77; "immersion" exhibit and, 60; zoo spaces and, 143; zoo text and, 80
Peircean Secondness, 77
Peircean Thirdness, 77–79
"people-movers," 120
Pepper 1930, 130
perception: by Surrealism, 136–37; external reality and, 177 n. 22; film theory and, 95, 100, 102–3; inductive statements and, 159; motionless objects and, 143; of art, 124, 134; of authority, 154; of images, 176 n. 2; of incomplete images, 147–48, 152; of paintings, 150–51; of photographed objects, 129, 134; of spaces, 144–47, 150, 176 n. 11; of stimuli, 102; sign and, 23; visual, 84, 134, 142–43; wandering viewpoint and, 84
Père David, 167 n. 33
Père David's deer, 47, 167 n. 33
permanent gaze, 118
perspective: 3D, 142; camera and, 103; cinematic, 101, 107, 117; fiction and, 81, 83, 86; hyperreality and, 19; invisible observer and, 95; panorama and, 113; photographic, 139; viewing animals and, 143; virtual gaze and, 107; virtual reality and, 50

pets, 29, 52, 153
Phantasmagoria, 113
Philipsbom's Department Store, *110*
phonetics, 69
photograph: as art, 133–35; as kitsch, 124; as object, 136; as objectified artifact, 135; as simulation, 42–43; contingency of, 125; definition of, 174 n. 2; distortion by, 131; emotional effects of, 124; exemplification of, 138; familiarity of, 136; forgery of, 134; illusion of consensus and, 143; memory and, 100; ontological nature of, 125; original object and, 60; postmodernism and, 125, 129; reality and, 128–29, 136, 138; representational thinking and, 59; semiotics and, 126–27; storytelling and, 100, 175 n. 22; the past and, 175 n. 15; unary, 139; vs. painting, 125, 133–34; wildlife, 135, 137–39
photographic seeing, 139
photography: as art, 124, 133; as static image, 95, 100; contribution to film by, 107; definition of, 174 n. 2; didactic value of, 124; fostering myths and, 138; iconism and, 32; invention of, 114; motion and, 92; original function of, 132; perception of reality and, 100; semiotics of, 126; signification of, 132; storytelling and, 128; Surrealism movement and, 136; theory of, 20; virtual gaze and, 107; vs. narrative, 128; vs. painting, 175 n. 1; wildlife, 126, 138, 140; zoos and, 21, 124
Picasso, P., 54, *55, 105*
Pierrot, 58
Plato, 14, 22, 41
Plato's cave, 147
Plato's *Forms*, 22
plot: definition of, 66; in cinema, 92, 100; in fiction, 81; in film theory, 103; in novel, 73–74, 77; in open text, 77; memory of, 73, 79, 172 n. 41; predictability of, 106; zoo text and, 86, 88. *See also* causality
Poetics of Prose, The, 66
poetry, 92
Polar Panorama, 117

postcentral sulcus, 104
postmodernism: animals in, 22; beauty and, 138; beginning of, 14, 164 n. 20; definition of, 164–65 n. 1; dimensional space in, 118; distinguishing features of, 13–15, 17, 129; environment and, 165 n. 27; film and, 92, 107; *Finnigans Wake* and, 170 n. 4; "hidden" Nature and, 136; hyperreality in, 17, 19; images in, 136; knowledge and, 16; linguistic code and, 76–77; narrative and, 59; painting and, 150; photograph in, 124–26; redundancy of, 43; shopping mall and, 109, 122; sign and, 37; space and, 19; spectacle and, 51; spectator and, 81; stories and, 81; subversion of reality and, 135; theater and zoos in, 92; *trompe l'oeil* and, 134, 139; vision and, 50; window shopping in, 120; writers in, 166 n. 17; zoos and, 13, 17, 49, 153
pragmatics, 71–72
Presbyterians, 75
preservation, 154, 161–62
press, 14
Pretty Woman, 106
Pribilof Islands, 167 n. 10
prison, 118, 174 n. 38, 176 n. 5
probability: in scientific hypothesis testing, 160; "third-order" simulation and, 42; truth and, 154, 157
production. *See* "second-order" simulation
protocinematic entertainments, 19, 107–8, 110–18, 120
Przewalski's horse, 48
Pudovkin, V. I., 95, 97
pun, 77, 176 n. 4

Radisson SAS Berlin Hotel, 50–51
rainforest exhibit, 15, 41, 57, 60, 62
rainforest, 41, 52, 62, 93
Rauschenberg, R., 134
reader: actualization by, 80–82, 98, 105, 170 n. 13; description and, 128; engagement of, 82, 88; explicitness and, 82; illusion and, 79; literature and, 78; semantic paradigm and, 70; semiotic progression and, 78; textual repertoire and, 83; textual strategy and, 83; writing and, 75
real time: fiction and, 79; in film, 93; theater and, 95, 99; zoo and, 91, 123
realism: cinema and, 17, 93; phenomenal experience and, 84; photographic, 128; reality and, 49; representation and, 43, 59; sense organs and, 14; vs. hyperreality, 59
reality: artifacts and, 54; Borges' map and, 37; cinematic, 100, 107; commodification and, 135; "crèche-ification" of, 57; denotation of, 100; description and, 72; diorama and, 116; dissimulation and, 61; fiction and, 78–81; fragments and, 45; "hidden," 128; hyperreality and, 17; illusion of, 100, 113; images and, 91–92; "immersion" exhibit and, 62–63; in art, 62; in Nature, 32; in postmodernism, 14–15, 129, 164–65 n. 1; meaning and, 82; memory and, 99; motion and, 141; naïve realism and, 147; "naturalistic" exhibit and, 55; nonfiction and, 72; perceived, 13; perception of, 44, 177 n. 22; photographic, 125, 127, 129, 131, 135–36, 138–39, 162; realism and, 49; representation and, 22, 147; representational thinking and, 58; semiotic, 43; shopping mall and, 50; sign and, 19, 23, 32–33, 36; simulation and, 41–42; Surrealism and, 136–37; synthetic statements and, 178 n. 17; theme parks and, 50; urban, 124; wildlife photography and, 138; zoo animals and, 93
referent: context and, 27; definition of, 24; denotation and, 84, 86; exemplification and, 138; imaginary animals and, 126; in art, 54; in fiction, 81; in film, 98; in language, 71; in Peircean Secondness, 77; in representation, 62–63; in semiotic progression, 78; in spectacle, 94; in

theme parks, 59; material and, 72; meaning and, 35; metaphor and, 86; Nature and, 65; nuclear content and, 30; photograph and, 126–28, 140; recognition and, 31, 34; representational thinking and, 56; semiotically real objects and, 26, 32; sign and, 25; signification and, *27;* "third-order" simulation and, 55

reflective judgment, 28, 31. *See also* determinant judgment; judgment

Rembrandt, 133

Renaissance, 40

representamen, 24, 28, 36. *See also* signifier

representation; by artifacts, 42; classification and, 31; denotation and, 62, 126; fragments and, 45–46; framed, 57; images and, 92; imagistic, 84; in Nature, 144; naïve realism and, 147; Nature and, 49; objects and, 126; photograph and, 126; picture and, 43, 126; reality and, 22, 147; resemblance and, 43, 126; semiotics and, 62; *trompe l'oeil* and, 62

representational thinking, 56–59

resemblance, 43, 62, 126

Rio de Janeiro, *57*

Robert's Rules of Order, 69

robotic dinosaurs, 171 n. 2

robotics, 42

Rokeby Venus, 134

Romanticism, 64

Rotterdam, 46

Rousseau, H., 49

Rulfo, J., 166 n. 17

Russians, 126

Said, E. W., 45

San Francisco, 49

Sant'Elia, 92

Santa Fe, New Mexico, *18,* 60

Santa Maria, 50

Sarnen Valley, 115

Save the Children Foundation, 58

Scale. *See* spatial dimension

schizophrenia, 15, 129

science: artifacts and, 45; as metaphor, 72; description and, 178 n. 26; zoos and, 20–21

scientific hypothesis: beauty and, 13; definition of, 155; zoos and, 162. *See also* hypothesis

scientific method, 154–55, 158

scientific theory: *ad hoc* statements, 156, 161; belief and, 157–59; criteria for, 155; definition of, 155; falsification of, 154–61; formulation of, 155; individuals and, 160; induction in, 158; observation and, 153, 156; paradigm and, 169 n. 7; prediction and, 156, 158, 160–61; properties of, *156;* scientific hypotheses and, 155, 157–58; universal statements and, 156, 159; verification and, 155–58; zoos and, 20

scopic dominance, 118

seahorse, 159, 178 n. 22

"second-order" simulation: artificiality, 64; beauty and, 64; definition of, 40–41; examples of, 42; laws and, 42; theme parks and, 54; zoos and, 49, 52–53

semantics: brain and, 169 n. 16; definition of, 71; in graphics, 72; of film, 98; of paradigms, 69; semiotics and, 74

Semiotic "primitive," 29

semiotically real objects: sign and, 26, *27,* 32–33, 35–36; taming and, 52; zoo animals as, 33, 44–45, 49, 84, 123, 136, 138, 162. *See also* objects

semiotics: actualization (signification) of, 32; definition of, 23; descriptive text and, 81; diegesis and, 171 n. 14; elements of, 24; event and, 32, 72; fiction and, 78; founding of, 164–65 n. 1; genes and, 58; language and, 98; model text and, 75; open text and, 77; process of, 34; progression of, 78; reality and, 32; referential text and, 80; representation and, 62; representational thinking and, 56–58; species labels and, 78; text and, 74, 84

semiotic triangle, 24
sensation: art and, 134; in postmodernism, 13; individualism and, 39; of objects, 36; sign and, 23
sentence: absence of narrative and, 86; as syntagm, 68; brain and, 169 n. 16; definition of, 67; empiricism and, 72; model text and, 75; pragmatics and, 72; sign and, 71, 79
shadow shows, 110, *111*, 148
ship of Theseus, 166 n. 6
Shklovsky, V., 134
shop window, 109, 110
shopping: commodity and, 108; hyperreality and, 122; shop window and, 110; vs. zoo spectatorship, 121
shopping mall: as kitsch, 50; cognitive acquisition at, 121; distortion of Nature and, 51; Gruen and, 122; merchandizing by, 64; modernist arcade and, 109; museums and, 64; postmodern, 122; sign and, 65; theme parks as, 122; virtual gaze and, 107
shopper's gaze, 108
Siberian tiger, 50
sight lines, 147
sign: artifact as, 45, 65; as denotation, 83, 86; as nuclear content, 30; classification of, 69; connotation and, 34; de Saussure's use of, 165 n. 3; definition of, 24, 84; descriptive text and, 81; examples of, 24; exemplification by, 138; in fiction and nonfiction, 72, 79; for Nature, 62; forms of, 67–68; hyperreality and, 17; images and, 128; imaginary animals and, 126; in art, 54; in film, 98; in narrative, 67; in representation, 62; in sentences, 68–69; in strings, 33; in theme parks, 55; interpretant and, 25; Kant's *schemata* and, 28; knowledge and, 23; meaning and, 34–35, 87; negative allusion and, 86; paradigms and, 69; Peirce's use of, 165 n. 3; Peircean Firstness and, 77; Peircean, 32, 67, 69, 77; photograph and, 126; placement in sentences, 70; reality and, 19, 23; referent and, 25; representation and, 36; schizophrenia and, 129;
self-signifying, 98; semiotic progression of, 78; semiotically real objects and, 26, 32–33, 35–36; signification and, *27*, 34; spectacle and, 94; types of, 25–26; use by zoos, 21, 43; wax museums and, 55; word order and, 71; zoo text and, 86–87; zoos and, 144
signification: by objects, 45–46, 55; by photographs, 126, 132; in denotative text, 84; knowledge and, 27; literature as, 79; meaning and, 25; of literary text, 170 n. 13; of Nature, 62; of sign (actualization), 32; overdetermined narrative and, 82; representational thinking and, 56; semiotically real objects and, 32; sign and interpretant and, 34–35; textual reality and, 82
signified, 24, 164–65 n. 1, 165 n. 3. *See also* interpretant
signifier, 24, 67–70, 164–65 n. 1, 165 n. 3. *See also* representamen
simile: as textual device, 67; falseness of, 88; in film, 98, 122; zoo text and, 80
simple (singular) statements, 158–60. *See also* universal statements
simulacrum: as sign, 43; definition of, 17, 37; fictive time and, 99; in art, 54; individualism and, 39; modernist esthetic and, 56; painting and, 133; photographed objects and, 127; in postmodernism, 13; representation and, 62; representational thinking and, 56; theme parks and, 55; wildness and, 52; zoos as, 42; zoos, 21, 144
simulacrums of simulation. *See* "third-order" simulation
simulate, 37
simulation: Baudrillard's taxonomy and, 40; conspecifics and, 17; definition of, 61; exhibit as diorama and, 56; framing and, 57; icon and, 25; in "naturalistic" exhibit, 43; in postmodernism, 14; in theme parks, 50; Nature and, 49, 63–64, 168 n. 21; of beauty, 65; of habitats, 65; of objects,

20–21, 37, 91; reality and, 41; representation and, 62; transcendence and, 64; zoo exhibit as, 45; zoos and, 59. *See also* dissimulation
Simulations, 166 n. 10
Singer, P., 174 n. 41
size-distance invariance hypothesis, 176 n. 11
Snooty (the manatee), 37–38
Snow, 14
social pastiche, 15
Sound and the Fury, The, 107
South American amphibians, 138
South Florida Museum, 37
Southdale, 122
space: appropriation of, 64–65; architectural, 64; constraints of, 100; dimensions in, 50–51, 148; enclosed, 145–46; entropic, 144; externalization of, 148; fragments and, 46; "immersion" exhibit and, 144; in film, 99, 101; in hyperreality, 145; in modernist theory, 144; in painting, 151; in postmodernism, 19; in virtual reality, 145; isovist and, 147; metaphorical, 97; motion and, 92; natural, 144; panorama and, 123; perception of, 144–48, 150; theatrical, 99; visual perspective of, 118, 142–43; zoos and, 144, 162
spatial dimensions, 118, 176 n. 2
species: as icon, 36; as individual, 29, 31, 38, 86, 165 n. 23; as remnant, 15; biological definition of, 160–61; endangered, 17, 20, 26; in taxonomy, 31; name of, 32; phenotypic characters of, 26, 35; reality and, 147; representational thinking and, 58; semiotic event and, 32; to be another, 61, 153; vs. biodiversity, 162
species label: as semiotic system, 33; as sign, 26, 28, 34, 36, 78; interpretation of, 22; signification of, 32, 35; truth value in, 72
spectacle: advertising and, 51; animals and, 17; definition of, 66; examples of, 66; expositions and, 121; football games and, 19; imitation and, 94; modernist, 101; my use of, 169 n. 3;

171 n. 13; theater and zoos as, 95; theme parks and, 122; voyeurism and, 44; zoos as, 21, 56, 66, 123
spectacular, 19
spectator: animal vs. image and, 91; cinematic format and, 92; cognition of cinema and, 102–3; collage and, 54; communication of art to, 98; definition of, 66; description and, 84; diorama and, 113–16; engagement of, 73; *Fantasmagorie* and, 112; film space and, 99; *flâneur* as, 107; future zoos and, 163; gaze of, 17, 44,123, 142, 150–51, 174 n. 41; hidden animals and, 97; hydraulic lift and, 116; hyperreality and, 49; "immersion" exhibit and, 60, 64–65; in literature, 56; interpretant and, 31; interpretation of zoo text by, 86, 88; movement of, 87; "omnipotent voyeurism" and, 119; painting and, 150; panorama and, 113; participation in film and, 97–98, 104–5; photograph and, 130–31, 136, 138; postmodern, 91,107; "reading" zoo exhibits, 80–81; representational thinking and, 58; sign and, 28–29, 32–36, 43; signification and, 32; spatial perception by, 145, 148; surveys of, 90; "thirty degree rule" and, 117; tourism and, 119–20; visual perspective of, 117–18; witnessing by, 19; zoo décor and, 95; zoo exhibit and, 117, 121, 123; zoo graphics and, 66–67, 72; zoos and, 21, 55, 153
Spectres, Les, 111
St. Paul's Cathedral, 116
Starship *Enterprise*, 38
statistics, 42, 157
Steele, D., 77
Steinberg, S., 20
Stellingen, 117
Still Life with Chair Caning, 54, 55
stories: as phenomenal experience, 84; as products of culture, 93; communication by, 58–59; events in, 106; examples of, 67; film theory and, 103, 105; fragments and, 144; in film, 94, 100; innateness of, 80; nar-

stories (*continued*)
  rative in, 66, 72; novel and, 73; reality and, 99; signification by, 69; syntagm in, 99; television commercials as, 91; zoo graphics and, 33, 66; zoo text and, 86, 88, 90
storytelling: as human, 88; as narrative, 73–74; camera and, 97; canonical format of, 106; definition of, 72; elements of, 106; in cinema, 105, 100; photograph and, 100; photography and, 128, 138, 175 n. 22; signification and, 32, 35; signs and semiotics and, 26, 28, 33–34, 36, 78; truth value of, 73; zoo graphics and, 90
subjectivism, 80
Surrealism, 136
Surrealists, 22, 136–37
Swift, J., 67
Switzerland, 114
symbol: as nuclear content, 30; as sign, 26; connotation and, 34; definition of, 25; *Fantasmagorie* and, 111; grammar as, 128; in photographs, 127; interpretation of, 73; language as, 78; linguistic code and, 170 n. 2; Peircean, 165 n. 8; Peircean Secondness and, 74; proposition and, 36; representation and, 62
symbol system, 25
synecdoche, 67, 98
synonym, 69, 86
syntagm: actualization of, 98; definition of, 67–69; in graphics, 72; in stories, 99; in television, 91; meaning and, 71
syntagmatic/paradigmatic relationship, *68*
syntax: definition of, 67; in film, 98, 172 n. 31; paradigm and, 69
synthetic statements, 178 n. 17

Tampa, Florida, 171 n. 4
taxonomic orders, 46
taxonomy: molar content of, 30; objects and, 34; recognition and, 29–30; referent and, 31; reflective judgment and, 31; representation and, 28

tele, 48
television (TV): ageographical city and, 167 n. 2; dimensional space and, 118; expositions and, 121; fictive time and, 99; history and, 14; sequential cut in, 91; theme parks and, 50; time and, 15; violence and, 19; wildness and, 52; zoos and, 43, 153
text: as enclosed space, 83; connotative, 84; denotative, 84, 89; descriptive, 81–82, 85; reader and, 82; repertoire of, 82–83. *See also* closed text; model text; open text
textual device, 67, 88
textual meaning (esthetic object), 81, 86
textual perspective, 81, 83, 86
textual repertoire, 82–83, 86, 88
textual strategy, 83, 86, 88
textuality, 66
Thailand, 61
theater: allographic nature of, 134; as mental instantiation, 134; contrast (architecture) and, 96; contribution to film and, 107; filmed, 97; *locus dramaticus* in, 96; means of imitation and, 94; mimesis and, 95; "naturalistic" exhibit as, 171 n. 16; perception of reality and, 100; similarity to zoo exhibit and, 95–96, 99; spatial constraint in, 100; vs. cinema, 98–99
Théâtre Ambigu-Comique, 114
theme park: as collage, 54–55; as shopping mall, 122; consumerism and, 15; hyperreality of, 50; referent in, 59
thermodynamics, 150
"third-order" simulation, 40–42, 55
This is Not a Pipe, 23
Thomas Cook, 174 n. 43
Three Musketeers, The, 73
Thumper (the rabbit), 74
thylacine, 136, *137*
time: camera lens and, 126; cinematic, 54, 123; closed text and, 92; fictive, 99; fragments and 46; motion and, 91–92; in modernist theory, 144; sign and, 17; space and, 148
tissue bank, 20

SUBJECT INDEX

Tobit Blind, 133
token, 28–30. *See also* type
tourism, 119–20
tourist's gaze, 107–8
transcendence: Disney's crocodile and, 50; in photographs, 127, 139; of art, 99; models and, 37; simulation and, 64
"transported immobility," 120
trompe l'oeil: as art, 134; as representation, 62; in panorama, 113–14; photography and, 139
truth assessment, 71
truth value: knowledge and, 72, 154–59, 169 n. 15; linguistic, 169 n. 15
Tursiops truncatus, 52
Twenty Marilyns, 16
Twin Towers, 42
type, 28–30. *See also* token

U.S. Marines, 175 n. 22
utilitarian philosophy, 174 n. 41
United States (U.S.), 38, 53, 113, 166 n. 17, 170 n. 2
universal statements: inductive inference and, 159; in science, 156, 158; in zoo graphics, 82, 87, 159. *See also* simple statements
unicorns: as icons, 25; degrees of "naturalness and," 21, 43, 92; semiotics of, 126
unseen seer, 118–19
Ulysses, 29, 73, 77, 106
"universal taming," 52
USA Today, 171 n. 2

Valéry, P., 65, 128
Valley of the Sarnen, The, 114
van Gogh, V., 73
Versailles, 163, 174 n. 38
video games, 14, 118
virtual gaze, 107, 110, 116, 120
virtual mobility, 107
virtual reality: as hyperreality, 42, 50; cinema and, 14–15; isovist concept in, 147; literature as, 55, 79–80; space in, 145
vision, 84, 103, 107
visual density, 148, *149*
visual repetition, 15

Wall Street, 160
wandering viewpoint, 83–84, 86
Wandering Violinist, Albony, Hungary, 140
Warhol, A., *16*
Washington, G., 113
West Edmonton Mall, 42, 50, 173 n. 16
West Indian manatee, 37, 160
Weston, E., *130*
whales, *53*
White Noise, 37
Wildlife Films, 92
wildness, 16, 52
Williams, J., 101
window displays, 110, 120, 122
window shopping, 50, 64, 120
winter gardens, 107
Wizard of Id, 76–77, 106
Woolf, V., 92
words: as signs, 24, 29; denotation by, 79; meaning of, 69; order of, 71; paradigm and, 169 n. 7; syntax and, 67
writing, 75, *90*, 134

Yosemite National Park, 48
Yucatán, México, *18*

zoo brochures, 137
zoo graphics: content of, 88–90; deconstruction of, 75; denotation in, 84, 89, 94; description in, 67, 83; descriptive, 80, 84–85; Eco's terminology applied, 165 n. 18; examples of, *85, 89*; facts in, 72; function of, 66; iconism and, 26; interpretant and, 31; interpretation and, 22; knowledge and, 27–28; linguistic code and, 170 n. 2; model reader and, 87, 90; nuclear content and, 30; photograph and, 138; recognition and, 29; science and, 153; semiotics and, 23, 33; storytelling and, 73, 91; textual device and, 67; textual strategy and, 83; universal statements in, 82, 87, 159
zoo guidebooks, 43, 120
zoological gardens. *See* zoos
zoological parks. *See* zoos
"zoomobiles," 120

zoos: absence of motion in, 44, 100; absence of narrative in, 44, 59, 106; Nature and, 20; perception of reality and, 100; arcade and, 123; artifacts and, 45, 47, 64; as artifacts, 42; as commodities, 64; as ecological fragments, 144; as service providers, 120–21; as spectacle, 19; Baudrillardean simulation and, 49; commodity experience and, 120; confinement and, 15; conformity of, 15; connotation at, 15; conservation vs. preservation and, 161–62; conservation and, 17, 121, 125; critique of, 20; fictive time and, 99; film esthetic and, 107; histories of, 164 n. 21; hyperreality and, 41; in modernism, 15, 59, *96*, 163; in postmodernism, 13, 20; in shopping malls, 122; other institutions and, 174 n. 38; panopticon and, 119, 174 n. 41; petting, 15; proselytizing by, 144, 162; protocinematic entertainments and, 107, 117, 120; saving animals and, 168 n. 34; science at, 153, 161–62; shop window and, 110; sign and, 144; space and, 100, 142–48, 150–51; spectator and, 44; survival and, 19; taxonomic orders and, 46; theme parks and, 50; uncontaminated Nature and, 60; use of photographs by, 124–25, 127, 129, 135–36, 138–39, 141; visual perspective and, 117; wax museums and, 52; window shopping and, 64, 120–21